Indie Rock 101

Indie Rock 101: Running, Recording, and Promoting Your Band

by
Richard Turgeon

Designs and illustrations by Richard Turgeon

www.indierock101.com

ELSEVIER

AMSTERDAM • BOSTON • HEIDELBERG • LONDON
NEW YORK • OXFORD • PARIS • SAN DIEGO
SAN FRANCISCO • SINGAPORE • SYDNEY • TOKYO

Focal Press is an imprint of Elsevier

Focal Press is an imprint of Elsevier
30 Corporate Drive, Suite 400, Burlington, MA 01803, USA
Linacre House, Jordan Hill, Oxford OX2 8DP, UK

Library of Congress Cataloging-in-Publication Data
Application submitted

British Library Cataloguing-in-Publication Data
A catalogue record for this book is available from the British Library.

ISBN: 978-0-240-81196-3

For information on all Focal Press publications
visit our website at www.elsevierdirect.com

09 10 11 12 5 4 3 2 1

Printed in the United States of America

Dedication

Music has been and will remain a lifelong journey for me. It's also a team sport. So I'll need to dedicate this book to several people—namely all the great musicians, engineers, and home recordists I've had the good fortune to work with and learn from since I was 13.

It's also dedicated to my mom for instilling in me an appreciation for words, and to my dad for his love of music.

Contents

PART 3 • Promoting Your Band

APPENDICES

Acknowledgements

Special thanks to my friend and recording partner in crime, Ron Guensche, for his invaluable suggestions and edits to the manuscript. Ron's a gifted engineer whose good taste, critical listening skills and deep technical knowledge I greatly admire and appreciate. He also plays a mean bass.

I'd like to thank some others who provided feedback and encouragement: David Battino, author of *The Art of Digital Music* (which you should check out) and Tim Pryde of San Francisco's Pebble Theory.

I'd like to thank my editor, Catharine Steers, for her enthusiasm, guidance and support in seeing the book through to publication, as well as my developmental editor Carlin Reagan. I'd also like to thank Focal Press for giving me the opportunity to share this book with you. And I give heartfelt and humble thanks to the incredible musicians (and their support teams and photographers) who granted permission to use their photos – I'm thrilled and honored to have you gracing these pages.

Finally I'd like to thank my girl, Becky Ruden, who lets me write and make music on nights and weekends, comes to my shows and sings along to my songs.

About the Author

Richard Turgeon got his first paying gig as the lead singer in a pop metal cover band at age 13, and was a musician even before that. Since 1993, he's independently released 6 records as a singer/songwriter (producing the last four) and self-managed his indie rock band, Johnstown. He's also played drums in a variety of bands, including San Francisco's The Tenderfew. You can keep up with his latest recordings and projects at www.myspace.com/richardturgeon

Introduction

WHAT'S AN INDIE ROCK BAND?

For the purpose of this book, an indie rock band is…

- Self-formed: The band, if it's not a solo artist, is formed by its members—not by a producer, manager or record label
- Self-sufficient: They write, record and (more so than ever today) produce their own material
- Self-managed: They book their own shows
- Independent: They are not signed to a major label

WHO THIS BOOK IS FOR AND WHAT IT WILL GIVE YOU

This book is for musicians at any level who want essential fundamentals on **Running**, **Recording**, and **Promoting** their band (getting their first few live shows and beyond). The book is divided into these three main sections that might follow this real-life sequence.

Part 1, Running Your Band, covers the elements most relevant to forming and running your band: the people, practice, and songwriting. Info on playing out and performance is in Part 3, but Part 1 is the section that covers more formative considerations around how to find good players, implement best practices and establish band dynamics.

Part 2, Recording, constitutes the bulk of the book, and it's chock full of information surrounding pre-production considerations, gear, and how-to basics, in addition to and instruction and techniques having to do with **recording**, **mixing**, and **mastering**.

Part 3, Promoting, covers the elements you'll need above and beyond the music to grow your fan base and define your "brand." Here, we'll cover graphic design fundamentals, your press kit, your websites and how to make a video.

The book is best suited for those who

- Are learning or can already sing and/or play at least one instrument
- Own or have access to a computer and feel comfortable teaching themselves at least simple recording software (like GarageBand) and hardware (like basic audio input devices)

There's a glossary of audio terminology at the end, so when you see a term in *italics,* you may want to refer to this glossary for a more complete definition or search for the term online.

Politically correct disclaimer: Indie rock isn't a boys club, so although you'll see the male pronoun or possessive used in the text, you can assume it's covering all you awesome female rockers out there, too.

BEFORE WE START...

We'll cover a lot in this book, but let's hit on some key points to keep in mind as we go and after you're well on your way:

Develop and trust your own ears, tastes, and instincts

Magazines, professionals, friends, peers and critics will all affect your opinions about your gear and music—or at least try to. Collaborating with others can be one of the great joys of music, and although it's certainly wise to seek feedback and advice along your musical journey, it's you who should remain the final authority on what to do, think, buy, or like.

You don't need to spend a lot of money to make music

There will always be "better" (i.e., more expensive) soundcards, computers, cables, hardware, and software to purchase, and they may very well be worth your money. They may even make your music sound better. But it's ultimately your ability, songs and dedication that will garner you an audience. You can play and produce music for a modest investment and still sound very good, so focus on developing yourself throughout your lifelong musical journey without getting too fixated on or indecisive about gear.

Protect your hearing

The body can recover from a good deal of abuse, but it can't recover from hearing loss. That's why it's critical for musicians, especially rock musicians, to protect their hearing with earplugs. Although plugs can definitely "muffle" or change the character of what you hear at practice, you'll get used to it after a few tries—and the muffling is a small price to pay for a lifetime of good hearing. Tissue paper or cotton is not effective in terms of preventing hearing loss, and foam plugs usually make music sound more unnatural or muffled than the following options.

The best kind of protection you can get are custom-made earplugs from an audiologist, which cost between $150-200. Explore other options by searching "musicians earplugs" online or by looking under "audiology" or "hearing" in the phone book. You should remold/replace custom plugs every year or two as the cartilage in your ears never stops growing.

A less expensive but still viable option is a pair of professional, *non*-custom plugs, which you can buy online for between $12-20 (Check out www.etymotic.com).

Now that we've covered some preliminary basics, we can move forward with some thoughts, techniques, and fundamentals for forming or running your existing band.

PART 1
Running Your Band

There's a certain magical quality to a great band. Any band essentially consists of people standing on a stage playing instruments, but a great band's collective chemistry, proficiency and energy inspire proportionate enthusiasm in their audience. As music is a universal language, the connection between performer and audience can be very powerful, even transcendent.

A great band is a matter of working with the right people, using practice time wisely and writing great songs. (All of this, of course, results in having a great live act, and we'll cover performance in Part III, Promoting). While a great band is usually the result of the members' collective experience and dynamism, there are certain basic principles that can help you make your band the best it can be, or at least better than it already is. Let's start by looking at your...

CHAPTER 1
People

Even if you've already formed your band, the following information is still pertinent to optimizing and running it most efficiently. And if you're just starting to look for people, let's start by covering the following sections.

HOW TO FIND GOOD BAND MEMBERS

Post ads online, hang flyers

There are plenty of places online to find other musicians to work with (www.craigslist.org is a no-brainer), but it can be just as effective—if not more so—to advertise with flyers.

The best place to post is in rehearsal spaces in your area because this is where real musicians go, the ones who are actually rehearsing, recording, and probably playing out. Other good spots to post include those where other musicians are known to hang out, like coffeehouses and bookstores. Below is a sample flyer with typical fundamentals and comments below each "field."

Headline (Who You Are And What You Want)
Attractive female indie rock-minded singer/songwriter seeks band.
Get to the point and make it sound interesting.

Influences (Who Do You Like, Sound Like, Want To Sound Like)
Veruca Salt, Breeders, Juliana Hatfield, Donnas, to name a few.
Try to limit this list to 4 to 5 artists. We all have eclectic tastes, but you don't have to list everything you like since you're trying to establish your overall vision here.

Links/Samples
Samples and photos at myspace.com/yourname
Official website: www.yourwebsite.com

I Have
A van, a rehearsal space, I know how to record, mix, and master on my computer.
Talk about what you have to offer.

Where/When
I live in the Mission District and am looking to practice twice a week in or near that area.
No one likes a long commute to band practice after work, and you won't either. Plus, GUDs (Geographically UnDesirables—anyone more than, say, 20–45 minutes away from the space) are often chronically late and eventually drop off. Save yourself some hassle and try to get someone in the neighborhood.

Age/Experience
I'm 22 and looking for people between 18 and 30. I've recorded and put out three records with different bands throughout college, and I've played at a good number of clubs in Philadelphia.
Setting expectations regarding your age and experience further helps you screen flakes and bad fits before you both waste time with an audition.

Goals
I want to do this for a living. Serious inquiries only, please.
A statement like this may sound a little strong, but you'll more likely hear from someone equally as ambitious if that's what you want. If not, be up front and say something a touch lighter like "I have a day job and plan on keeping it—just looking to jam locally on weekends and do the occasional local show."

For Consideration
Please send links to MP3s and preferably, photo(s).
As you get more experienced—or you're paying for your rehearsal room time—you'll want to screen with samples and, yes, even photos (especially for lead singers) before you potentially waste time auditioning them. At least get a recording—any recording: it's still the best way to judge people's level of experience and tastes without meeting them.

FIGURE 1.1
This is one of the best places to find your next band: flyer boards at the local rehearsal studios. (Photo by R. Turgeon).

Network

In business and in the music business, there's no faster way to fill a post than through referrals (who also come with at least one reference right off the bat). If you're just starting out or are new to town, go to local shows, start chatting with good bands, and pass out your music. Don't be shy: ask if they know anyone who's looking for someone like you to play with. Every local music scene features a smaller cast of characters than you think. So start passing out demos to bands you like or want to sound like. If you're polite and have good material, most people will be happy to help you. (More on networking with other bands as it pertains to live shows will be discussed in "Playing Out," Chapter 12.)

A word on appearance

It would be nice to think that bands are judged solely on the merit of their music, but in the music biz, that's never really been the case. Some might say that it's superficial to concern oneself with one's band's image or appearance, but whether it's fair or not, doing so gives you an advantage. Even if the rest of the band doesn't want to amp it up out of some misguided sense of purity or comfort, it's important that at least the singer projects their band's ethos, music, and image by carefully considering what they wear on stage. It's less important that your act try to dress cool than it is to at least discuss how you'll be presenting yourselves—and that there's self-awareness cohesion and competence in this area. Regardless of what your image (or non-image) may be, a useful exercise is to look at video or rough photos of your stage presentation and ask yourself a few questions: Do we look like a unit? Do we somehow stand out from the millions of other bands out there? Would ours be a band someone might pay to see or hear based solely on our photo?

As you start to meet and audition people, you'll also want to be sure to:

Find like-minded players

It's said that the things that bind and break couples are sex, money, values, and beliefs. When mates are aligned on these things, the relationship has a better chance of lasting. It's not surprising that similar considerations might be made in choosing band mates. More specifically:

- Do you like the same bands and types of music?
- Do you have the same goals for the band?
- Do you have similar personal values and beliefs? (drugs, work ethic, punctuality)
- Are you clear on (and happy about) roles and responsibilities?

Collaborating with diverse opinions and personalities can be rewarding and result in some great music, but you ultimately want to be able to say at least a soft "yes" to these questions. As you gain more experience in your band and possibly in other bands, you should be able to more quickly and instinctively evaluate whom you should associate with or avoid.

Once things start to firm up a bit, or it's time to narrow down to a few choice candidates, it's important to keep a few things in mind…

No weak links

When choosing your bandmates, it's important that everyone adds value and that there are no obvious weak links performance or business-wise. People who are playing at the same level but who can also push and challenge each other make a good team. Ideally, you want to choose people who are happy in their respective roles and who work well with others—a valuable, developed skill in itself that plenty of good musicians never develop.

The lead singer is, of course, the band member whose primary role is to initiate and maintain that connection to the audience, so they're naturally the most prominent member of any collectively decent band. In the minds of general audiences, the singer sets the level for the band as a whole. While it's important for every member to be good, it's important to keep in mind that a band even with the best musicians will rarely rise above the charisma and ability—or lack thereof—of their singer.

Regarding related roles and how they relate to the logistics and business around running a band, it's most effective when roles are clearly defined and everyone contributes. The bands that tend to get ahead fast are those who are clear on what their role is as both a performer and as part of the business, if they decide to treat the band as such. Individual members may handle PR (getting reviews, sending out press releases, and setting up interviews), getting airplay, booking, design/photography/website, distribution, fan relations, and more. If you're looking to earn money or a living as a musician, you can move mountains by adopting a business mindset and dividing and conquering management tasks with your team.

When to cut an audition short

Even if you did screen a player beforehand, sometimes it'll be obvious to you that it's not a fit after just a few minutes. If that's the case, politely tell them you have a good idea of what they can do and thanks for coming out. Even if you don't have another candidate coming in for the next 20 or 30 minutes, you and the band can put the time to better use by practicing without the candidate.

Every team needs a captain

This point is related to the last about everyone being clear on roles. Just because a group of musicians form a band doesn't always mean that the band is a democracy, or that it should be. Most bands are not bona fide democracies unless, as was the case with The Doors, that's the arrangement agreed upon by all band members. Many bands would save themselves a good deal of time by being open and honest about who makes final decisions and how they're made. It's often best if and when a band has a clear leader or two since having some sort of hierarchy makes any organization more effective, especially when its leadership is effective. Music made by pure consensus can also be in danger of sounding watered down.

There have been exceptions to this paradigm, but the bandleader or leaders tend to be the lead singer, lyricist/songwriter, and musical arranger. A typical leadership paradigm in rock is a pair of individuals like a lead singer who writes the lyrics to the riffs or songs written by a lead guitar player and/or multi-instrumentalist "musical arranger." What's surprising about most bands is how talented and multi-faceted most individual members are, so it's not surprising that when collaborating, their collective efforts far exceed the sum of their parts from both a music and business standpoint. Regardless of who's running the show and who's most invested in the band's success, it's important to keep in mind that the captain's job is not to do everything himself or be a one-man show—it's to bring out the best in their team and direct their efforts.

NEXT, CONSIDER...

By now you may have met some good people and decided it's a good fit. You could wait until you've played together for a bit longer, or until you're ready to promote yourself, but it might be time to start thinking about...

Your band name

It's good to make sure that your friends and band mates are happy with any band name, but it's even more important to make sure that it's not already being used or legally registered by another band. You could end up losing a lot of money and time promoting yourself under a shared name you might ultimately be forced to give up or even be sued for.

You can avoid this by taking a few simple precautions like Googling your top picks online and searching them on music sites (like allmusic.com) and digital distributors like iTunes Store, MySpace, and CDBaby (more on distribution in "Part 3—Promoting Your Band"). If you're serious about selling your music, you might want to pay about $300–$500 to register your band name as a trademark. You can get the ball rolling yourself by visiting the United States Patent and Trademark Office (www.uspto.gov), or by paying a lawyer to help you with this process. A list of beginning and established lawyers looking for new clients can be found on CDBaby (www.cdbaby.net/picks-lawyers).

Know when to part ways

If a member doesn't seem to fit in or consistently act as a positive influence, ask him why and take action from that point. Never be too hasty in making this decision, but don't hesitate to take action if the band as a whole is suffering. Bands can consist of both personal and business relationships, but you'll know when to part ways if you must. If you do need to fire a band member, and you're the band's leader, afford the one being ousted (and yourself) some modicum of respect by at least calling him or, preferably, meeting him in person and breaking the news yourself.

When you get the right combination of people together and it rocks, you'll know it. Then you'll be set to regularly…

CHAPTER 2
Practice

There's probably no aspect to your music life that's more important than regular practice. It's what separates mediocre bands from the ones that get noticed and eventually—if they stick together—get paid. Being in a band and practicing should be fun, but if you're serious about getting good or better, it's best to be disciplined, dedicated, and take the following into consideration. The first thing you'll need to do is...

DECIDE ON A PRACTICE LOCATION

Practice anywhere you can, but before deciding on the location, you should know the pros and cons of having a "funded" versus "found" rehearsal space. A funded space is simply one that you and/or the band pays for, while a found space is one that a band member or members is personally invested in or connected to (like their parents' basement or garage), meaning they can provide the practice time for free.

With a *funded space*, you don't have to make compromises on personnel who have something to offer above and beyond their musicianship. Paying for time also helps keep things professional, where your time and efforts are focused and suffer from fewer distractions like parents and friends dropping in and any limiting conditions you may face on someone's "found" turf (Mom or neighbors want amps off at 10 p.m. sharp, etc.) Plus, most rented facilities are soundproofed and they may have a store if you need food and gear. In other words, they're designed to accommodate practicing musicians.

It's worth mentioning a distinction between the hourly and so-called "lockout" (since it's yours to collectively rent, lock, and unlock) funded spaces. Hourly spaces are great for auditions and projects you know will be short-lived (like if you're rehearsing for a single or just a few shows), but most serious bands prefer a lockout space since it allows them to keep their gear set up, whether it's for routine practice or recording (Figure 2.1).

The main advantage of practicing in a *found space* like your drummer's garage is fairly self-evident: It's free. It just makes it very difficult to work on the *band's* schedule or, if you need to, fire the drummer—after all, it's his garage.

Whichever route you decide to go, the important thing is that you…

FIGURE 2.1
The Tenderfew's Chip Dalby surrounded by elements common to most lockout practice spaces: empty beer bottles, random stickers and flyers, tons of gear, and of course, holiday lights (Photo by R. Turgeon).

PRACTICE AS A BAND REGULARLY

Try to get together at least twice a week for at least 2–3 hours, preferably without excessive drinking, drugs or other distractions. The more you practice, the tighter you'll sound—it's that simple. At first, you'll need to use both weekly practices to get a 10–12-song set tight. After everyone knows the tunes and there are no personnel setbacks, you can start "staggering" practices, meaning you use maybe one practice every week to keep your "core" set (the 10–12 you've already learned) sounding tight, with the remaining practice dedicated to learning new material. Or use the first hour of practice for your core set, the second hour for new material.

Another thing to think about is allotting some time at the end of every practice to discuss band strategy, logistics, and business. If you're serious about moving the band forward, there should be plenty to do and discuss, so it makes sense to have band meetings right after practice. Of course the shop-talk doesn't always have to be in your rehearsal space; sometimes taking the meetings offsite to the local coffeeshop or pub can be a nice change of scenery while boosting band morale and bonding.

A word about scheduling practices

You can avoid a lot of scheduling headaches by asking your band to commit to regular timeslots. If you don't want or need the band to keep two slots a week clear—or you only want to practice one night a week—establish one weekly slot as your "main" night and the second as a "reserve." That way, if someone can't make the main, you have the reserve already on the calendar. If someone keeps flaking or outright doesn't seem capable of respecting these slots (i.e. everyone else's time), find someone who can.

"JAMMING" VERSUS "PRACTICE"

'Jamming' can be a term defining, from its most general definition to more specific:

- Musicians getting together to practice or play for their own sake, with no particular goal in mind other than to have a good time.
- Musicians getting together for a non-committed "getting to know you" session (usually not a bona fide audition but rather when musicians are feeling out whether or not they want to work together).
- Improvising around a new groove or song created on the spot, improvising within the context of a song the band already knows, or improving within the context of a *cover song* (a term describing a song made popular by another artist).

It happens, but jamming shouldn't be a term used to describe:

- *Audition*: That commitment-free "first date"/audition meeting to see if it's a good fit, sometimes held at practices of already-formed bands.
- *Practice*: When already formed bands get together at regular time slots to rehearse their material.

If you have any ambition to be in a professional or even decent band, you need to make sure your band members understands these distinctions and are on the same page about how your practices are structured around them. To that end, it's helpful to be clear on why you're getting together from the outset and especially as you reconvene. Is it fun for its own sake, or instituted to reach a goal like playing out at a professional venue, securing a record deal or practicing to record your first CD? Be up front about your goals and what practices and your music mean to you so, your practice time is well spent and focused, whether that includes more improvisational jam-time or not.

FORM A SET LIST

This is jumping a bit ahead to the "Playing Out" section in Chapter 12 of the book, but once your band starts learning the songs, you'll want to start putting them in an order that optimizes your live show. It's also important to note that the song order for any band's live set is often quite different from the song sequence on their record, a fact that helps to pronounce the difference between the live and private listening experience. Regardless of the genre they're working in, filmmakers typically put the "best" three scenes in the beginning, middle, and end, and you'll want to structure your set list similarly—at least in a general sense. A good rule of the thumb is to open and close with a bang, and keep any slower, less dynamic or newer material in between. In addition to setting the mood and pace of your show, practicing your set list well in advance will help the band perfect its starts, stops, and segues between songs. If do you a *cover song* or two, it's always helpful to put them in the beginning, middle or end of the set when you might need to pull the audience back in with the recognition factor. Minding these details can make the difference between being perceived as a bar band that goofs off too much during and between songs and one that blows people away with a tightly orchestrated and well rehearsed rock show (Figure 2.2).

Another thing to consider when putting together your set list is your band's gear. There's nothing more irksome to an audience than when a band spends too much time between songs swapping out guitars, tuning

FIGURE 2.2
Playing new material live is one of the best ways to tell what songs are keepers (or not). Juliana Hatfield on this photo: “That night I played a brand new song that I had just written and no one had ever heard before. The crowd seemed to really like it and clapped just as loudly, if not more loudly, for that unknown song as they did for all their familiar favorites.” (Photo courtesy Rick Marino)

up, and the like. It's acceptable for a guitarist to use different guitars on different songs, but make sure that it's not more than once or twice a set if it can be helped. Ensure that your set "clusters" your songs by what guitar he'll be using and ask him to do it as quickly as possible to avoid a lot of dead air onstage. Speaking of quickly, unknown bands should try to limit their sets to 35–45 minutes—about 8 to 10 songs max. You can always do the 2-hour stadium sets when you're playing there, and remember, when it comes to showbiz, it's always better to leave them wanting more.

CHAPTER 3
Songwriting

While rock songs are typically simple in composition and arrangement, a classic song with a great hook and lyrics is more difficult to write than one might think. Most musicians who go on to become accomplished songwriters have an interest and background in the art and craft of writing itself and understand the written word's relationship to music; they generally don't need much, if any, academic instruction in songwriting. That said, there are some general guidelines that can help inform even those who are better suited to focusing on other areas like their instrument, directing videos, or mixing.

Even if you're not that great of a songwriter or have no desire to write songs, it still helps to know songwriting basics so you can interpret, collaborate, and write your part in an informed manner and in the proper context of the song and your band. Even if you're not the person who brings the lyrics, chords, and melodies to the table, being able to recognize and work with song structure is a valuable skill in itself. There are a lot of songwriters and beginning bands who may have great hooks, but maybe their songs sound loose, unfocused, and inconsistent. This is where experienced producers often help out, many of the great ones being excellent songwriters and arrangers themselves.

But before we get into some techniques, definitions, and philosophies that surround the art and craft of songwriting, let's cover...

GROUP INTERPRETATIONS OF SONGS

Whether your songs are written by a single member at home or between band members at practice, band practice is where they're individually interpreted and arranged or brought to life. Although we've been putting an emphasis on discipline, the exercise of fleshing out each song can be a very creative process, even if there's only one main songwriter bringing material to the table. There's a lot of potential for egos to collide here, so try to remain as open-minded and diplomatic as possible with your mates during the creative process (Figure 3.1).

It helps to be self-aware in terms of your place in the band. Are you the main songwriter, a songwriting "partner", or just a sideman? It's always good to grant the songwriter any sort of veto power in terms of hard "no"s on how a song will be played, and if they're so inflexible that you can't have an opinion, then they're looking for a session player or sideman (who normally, by definition, gets paid more than the social life that comes with being in a band). If you're not the songwriter or start finding that you don't have much room to make creative contributions to the band, either rethink your role in general ("Should I write my own songs

FIGURE 3.1
With San Franciso's Birdmonster, the songwriting process is never quite the same twice. The band wrote most of the songs for their second full-length album From the Mountain to the Sea in a cabin near the Mojave Desert. (Photo courtesy Greg Crane. www.gregcranephoto.com)

or stay a drummer?") or look for a project that gives you the degree of "say" you're looking for. Also be realistic and recognize that it's tough to drive or maintain much creative control at an indie band level if you're not the main songwriter/musical arranger (and again, these two often go hand-in-hand, even if that person is not writing the lyrics).

But enough about diplomacy and collaboration. It's now time to get down to...

SONG STRUCTURE BASICS

A song's *time signature* looks like a *fraction*, but it's actually the number of notes per measure over the *kind* of note (quarter note, eighth note, etc.). Most rock songs you hear on the radio, including dance music, are in 4/4.

Notice how a song's dynamic and structure often change or are dynamically emphasized in *four-measure* cycles. As an example, try composing a simple song and making changes like introducing a cymbal crash, or a new second guitar or synth part, every third or fifth measure. Sounds odd, doesn't it? We're all used to hearing it, but it's important to know that most rock and techno songs alike usually build or are dynamically "marked" in four-measure cycles.

A typical pop or pop-rock song (yes, with radio-friendly "indie" and "alternative" rock falling into this category) is often structured like:

- *Brief intro* (2–4 bars): Often introduces a guitar riff.
- *First verse:* Usually 4–8 lines that rhyme in an "ABAB" fashion, that is, the end of the first line rhymes with the third, the second with the fourth. (Of course, there are variations on this typical rhyming scheme.)
- *Chorus:* The chorus can consist of the same or, more typically, different chords than the verse. Typically, the choruses appear after each verse and, unlike the verses, have the same lyrics. The chorus is lyrically and musically the main refrain of the song, and oftentimes where you'll find the song's "hook," or most memorable melody.
- *Dynamics:* These describe a particular song's or arrangement's "ups and downs," or "peaks and valleys"—that is, the difference between the soft and loud sections. Dynamically speaking, the chorus is generally louder and more lush in its arrangement than the verse, possibly introducing new elements like a second guitar or, more commonly, background vocals that harmonize with the main melody.

It's also fairly typical to keep building on successive choruses with additional tracks and/or more complex arrangements to add interest and intensity as the song progresses. "Don't bore us, get to the chorus" is an old saying that can immediately be illustrated by listening to any number of 1980s pop songs, where the choruses sound very big and bombastic both in terms of arrangement and relative volume in comparison to the verses.

- *Second verse:* If you're song has "narrative" lyrics, that is, it's telling a story, amp it up a bit here, building on your first verse to carry the story a bit further. More instruments, sounds or flourishes are usually introduced at this point to build on the arrangement and hold the listeners' interest.
- *The bridge*: Or the so-called "middle eight" provides a break from the verse-chorus structure by introducing some variation in the form of a new melody and chords, often in the same or a related key. Typically, this section—often clocking in around the middle or "two-thirds" mark of the song—is a musical detour of sorts but not a huge departure from what you've heard so far. A bridge/solo that veers pretty far from the verse/chorus might be called a "breakdown." A bridge is often followed by a solo of some sort—a guitar solo being the kind that rocks hardest, of course.
- *Chorus, chorus, fade…:* After the bridge and/or solo, pop songs typically go right to the chorus once more, and possibly repeat the chorus twice. As mentioned earlier, try building on your choruses by throwing in an additional, cool new lead guitar line, countermelody, background vocal, etc. That's not to say throw *all* that stuff in, just try to make each successive chorus a bit more lush or somehow distinct from the last.

A few more songwriting fundamentals and suggestions to keep in mind as you develop your own material:

TELL A STORY

Poetry and impressionistic lyrics are hallmarks of rock and roll, but if you think about it, most classic, timeless songs tell a story.

Recall just one of your favorite songs of all time. By its nature, pop music is compositionally and structurally simple music (beauty and simplicity often going hand in hand). Put the lyrics in front of you and read them in time with the song; this always drives them home in a way that reading them "dry" without the music can't. Notice how great, truly literary songwriters like Kris Kristofferson, Bruce Springsteen, and Sting can tell a story in about four lines? Listen to/read "A Boy Named Sue" or "Sunday Morning Coming Down," "The River" and "Don't Stand So Close to Me," even if it's just the first few lines. They're impressionistic and evocative enough on the page, but when performed by the singer, notice how you can practically see the narrative unfold in time with the song. They evoke a mood and paint a picture in just a few bold strokes.

The lesson here is to think cinematically. In writing song lyrics, it helps to study the related discipline of screenwriting. Screenplays need to tell stories *visually* and in the least amount of words possible. A great introduction to the art and craft of screenwriting is a book called *Screenwriting: The Art and Craft of Writing for Film and Television* by Richard Walter. Read a few good screenplays of movies you like and are familiar with and notice how very few words are used to paint the scene and move the story. Exercise the same economy and discipline in your lyrics. (*Note*: You can order screenplays online or look for a more limited selection at your local bookstore.)

AVOID PUNS, CLICHÉS, AND STOCK PHRASES

When you see cringe-inducing advertising in print, it's usually a result of a cliché or pun—verbiage that's trite, hackneyed, lame. There's something to be said about using common, everyday language in rock as rock is a form of folk music—you just want to do so in a way that's fresh, unique, and original.

Politicians and some contemporary country music songwriters may love 'em, but in good writing, avoid stock phrases, puns, and clichés.

WHAT MAKES A GOOD SONG?

Obviously a subjective stew, but there are some common denominators:

- *Originality*: It's OK to sound like your heroes when you first start out, but keep learning and growing so you start recording and performing songs that sound unique to *you*.
- *Passion*: One only has to browse the CD kiosk at Starbucks to find a lot of music that's bland, safe, and downright boring. Rock music should not be generic, tame, lame, polite, demure… *nice*. No, great rock—and all great art and great lives—is borne of passion.
- *Melody*: Repetition is a cornerstone of popular music (verse, chorus, verse), but even in conventionally structured songs you'll want to try to keep your melodies interesting and varied (this is where having a great lead singer and musicians helps).

- *Simplicity*: It's usually cooler and classier to err on the side of simplicity. Not dumbed down, just simple. Think about it: most hooky, eminently hummable hit songs are pretty simple in terms of their structure and arrangement.
- *Humor:* Irony + wit + smart + funny = *good*. Pedantic, sophomoric, sexist, preachy, overwrought, maudlin, trite, stock = *bad*.

PROTECT YOUR WORK

Even if you don't intend to sell your songs or record, it's to your advantage to register your music with the U.S. Copyright office. See the 'Sound Recordings' link for the appropriate forms and the 'inducements and advantages' link on this page for more information: www.copyright.gov/register (> Sound Recordings).

Now that you've formed a cool band and know how to write a decent song, it's time to think about recording a demo, because that's what you need to start scoring gigs and generating a buzz—both of which will usually help you build your fan base.

Practice makes perfect indeed

Some of the greatest producers in the world put a lot of emphasis on the band rehearsing very thoroughly before they even think about recording. Firstly and most obviously, time is money and the studio is not the place to work out arrangements or conduct freewheeling musical experiments. Practice makes perfect, and that collective, best-effort perfection is what you want to walk away from the studio with in your hand after the bill is paid.

Speaking of money, if you're not a solo act and everyone's agreed to chip in for the recording, make sure you collect it before the recording is complete. You will very likely have to pay any studio bill before you get your masters, and you won't want to wait for them or cover the difference yourself for anyone backing out at the last minute.

ARE YOU READY TO RECORD?

This question is ultimately up to you and any collaborators or band mates, but there are some considerations that can help you make that decision. Some bands start recording very soon after they form. Others practice for months before even thinking about recording. Most bands seem to sit somewhere in the middle, where they start thinking about at least recording a demo after choosing 3–4 of their best songs from a wider set of perhaps 10 or 12 the band has learned.

In the next section, "Part II—Recording," we'll explore considerations around why you're recording, who should record you, and where and how you'll record; then we'll cover gear, the recording process, mixing, and mastering. Answering these preliminary considerations and having this fundamental production information could put you a few years ahead in terms of need-to-know basics and help you avoid a lot of lost time and money in the studio.

And now, onto…

PART 2
Recording

Now that your band is formed, you've got a good set of songs together, and you've been practicing for a while, it's time to record a demo. This section is divided into four sections: Pre-production Considerations, Gear, Recording, and Mixing.

CHAPTER 4
Pre-production Considerations

COMMON RECORDING APPROACHES

When it comes to the production process and gear, everyone has different styles, processes, and preferences. Some bands prefer to record all at once for that live energy and feel. Other bands and solo artists prefer to build recordings from the ground up, typically laying down rhythm tracks first (drums, then bass), then guitars and keys, and lastly, vocals.

With the advent of digital audio and the Internet, more bands and artists are recording virtually or remotely, which can entail Person A posting tracks online for Person B and others to record tracks on top of the foundation, building until the recording is finished. This is a common way to work in more "studio-oriented" (vs. band) productions and genres like dance, pop, rap, and hip-hop, where the music is often produced before the "star" vocalist's tracks are laid down.

Another approach is to use *session players*—basically musicians for hire that play on records for a main artist or band other than their own, if they even have their own project. For example, the vast majority of country music is recorded in Nashville, where it's common for very experienced session players to record either before or along with the main act, learning and often sight-reading their parts from charts (sheet music) on the spot.

Work-for-hire agreements for session players

Let's say you're a solo artist (producer, songwriter, performer, or all three) who just wants to record another performer (or performers) on your record, —but the two of you are not in a band and you want to retain and protect your legal claim to sole authorship and the copyright. This is where a *work-for-hire agreement* comes in handy. They're also called Session Player or Sideman Agreements, and you can find plenty of leads online by Googling "music contracts."

While you'll see that there are plenty of template-style contracts for sale online and in bookstores, a better route might be to have a lawyer tailor the contract to your needs for a reasonably modest fee. As we previously mentioned in the 'Your band name' section, a list of beginning and established lawyers looking for new clients can be found on CDBaby here: http://cdbaby.net/picks-lawyers. You can also find one by searching your local bar association.

Remember that when it comes down to business, it always helps to remember to negotiate or at least ask for a break: when you (briefly) email prospective entertainment lawyers about your musical activities and needs, ask if they can work with you on their rates. You may be surprised at some of the flexibility you can get when you just ask. (Yes, even with lawyers.)

You can record any number of ways, and eventually you'll discover your own preferences and methods based on available time, resources, band mates, available gear, and other considerations.

All that said, most artists record one of two ways, at least in a general sense:

a. They record their parts together as a **live band**, then go back and *overdub* their parts, meaning they re-record the entire track as a group, re-record an individual track or tracks on a whole song (like a vocal take), and/or *punch in* only at certain tracks and/or areas of the song to correct specific mistakes (like a flubbed bass note or sharp vocal line).

b. They start with **scratch tracks**—tracks that might ultimately be "scratched" or erased and replaced with better tracks—and build from there. Typically, the process starts with the tempo being established by a **click** or guide track, which is simply a track that "clicks" like a metronome for the band to record to (more on this ahead). Again, typically, the singer and at least one instrument like guitar or keys (if they're not played by the same person) will record their scratch tracks to a click track as foundations for the drummer and the rest of the band to build on.

There are certainly advantages to both approaches. A lot of bands like to capture the energy and feel of a "live" performance, with the band recording their parts together. If the band is well rehearsed, this can also potentially save a lot of time.

The scratch track approach is of course the way to go for solo artists, or artists working remotely and/or sequentially (i.e., not at the same time and location). Another advantage to the scratch track approach is that the scratch track or tracks can be set to a click track, which helps keep the tempo more consistent. Also, once scratch tracks are done—even if it's just a click, the singer, and a single "scratch" guitar track—that person doesn't have to repeat their performance over and over for an instrumentalist (like a drummer, for instance) that might not be getting their part right.

Along with any logistical or personal preferences, in choosing a recording process it also helps to ask yourself…

WHY DO I WANT TO RECORD?

As you'll see, asking this question helps answer the "Where should I record?" question, addressed in more detail below.

You could have lots of reason for wanting to record. Maybe it's just for fun, or perhaps just to teach a band mate a new song or get it down for your own reference. Maybe you want to learn more about the recording process, and that's always time well spent. But none of these reasons are for "public purposes."

Your demo is "public" even if you're only using it in the hopes of recruiting other band members. You should want to work with the best band mates you can find, and in the recruiting process, they'll likely gauge your overall level of experience by the production quality of your demo as much as by the songs and talent that went into it.

Since the aim of this book is to "get you out of the garage" and onto the path of playing out live and sending recordings to radio stations, online channels, etc. (i.e., becoming "public"), your demo needs to sound great—it's the reason for recording that we'll focus on. A demo needs to be more than just a calling card, or resume—it's your art and commerce. If it's good, it can open doors for you. When you give a semi-pro or pro in the music business an amateurish-sounding demo, they'll know it in about two seconds—especially in an age when anyone with GarageBand, some decent skills, and a good ear can make a very nice sound recording.

Don't shortchange yourself—make your first public demo count.

WHO SHOULD RECORD US?

You have three choices:

- The indie, D.I.Y. ("Do It Yourself") route
- A project or home studio (i.e., you are paying for someone else's time and location) or
- A professional studio with a producer/engineer

As recently as a few years ago, you had to pay a high hourly rate to record in a well equipped recording studio stuffed with gear worth tens to hundreds of thousands of dollars. Then digital recording came along and changed everything: most of that pricey, cumbersome analog gear and tape were replicated by computer hardware and software, and for a lot less money. The advent of digital audio has literally put hundreds of studios out of business, and while that's bad news for them, it's great news for you.

The fact that recording has become so accessible to so many people means that even if you choose to make your first recordings in someone else's studio, you can find someone who's really good for about $15 an hour—even in big cities like New York, Los Angeles, and San Francisco. In fact, with a larger number of skilled producers and musicians in those big cities competing for the same business, it's almost easier to find low, competitive hourly project studio rates using community websites like craigslist.org than it is in the smaller markets (If you try craigslist, click the 'community,' > 'musicians' link and search 'record' and/or 'studio').

While we'll cover the audio and *production** basics you should know—along with a lot of simple tips and tricks—there are some good reasons to consider recording your first project (or first several) with a good producer/engineer. There are also advantages and disadvantages to doing it yourself, indie-style, and going with a low-rate project or home studio.

Let's look at these three categorical choices for you to consider, along with the possible pros and cons of each.

*An umbrella term we'll use to encompass recording, mixing, and mastering.

D.I.Y.

If you want to try your hand at recording your own demos, you'll need some gear (covered ahead). But the good news is that building your own project studio (or "DAW"—Digital Audio Workstation—is relatively inexpensive, and production (as opposed to playing and performance) is a fascinating world and a skill set you can use and enjoy for the rest of your life (Figure 4.1).

Pros: D.I.Y. builds your skill set and "ear." Being able to produce empowers you. It's fun if you have an appetite and aptitude for it. After some initial investments, it lets you record and release records at your own pace and without breaking the bank, whether it's your own material or someone else's.

Cons: Producing yourself makes it hard to be objective. You might not be ready to record your band. You may encounter potentially steep learning curve(s) if you're not tech-savvy, and even if you are, you have to stay current with available hardware and software from both a knowledge and financial standpoint. If you want to record a "live" band (everybody plays/records together) or drums, you'll need multiple mics and a soundcard that can accommodate all those channels/mics (more on gear ahead), which can get pricey. You might be better off recording drums and everybody at once at a project or home studio.

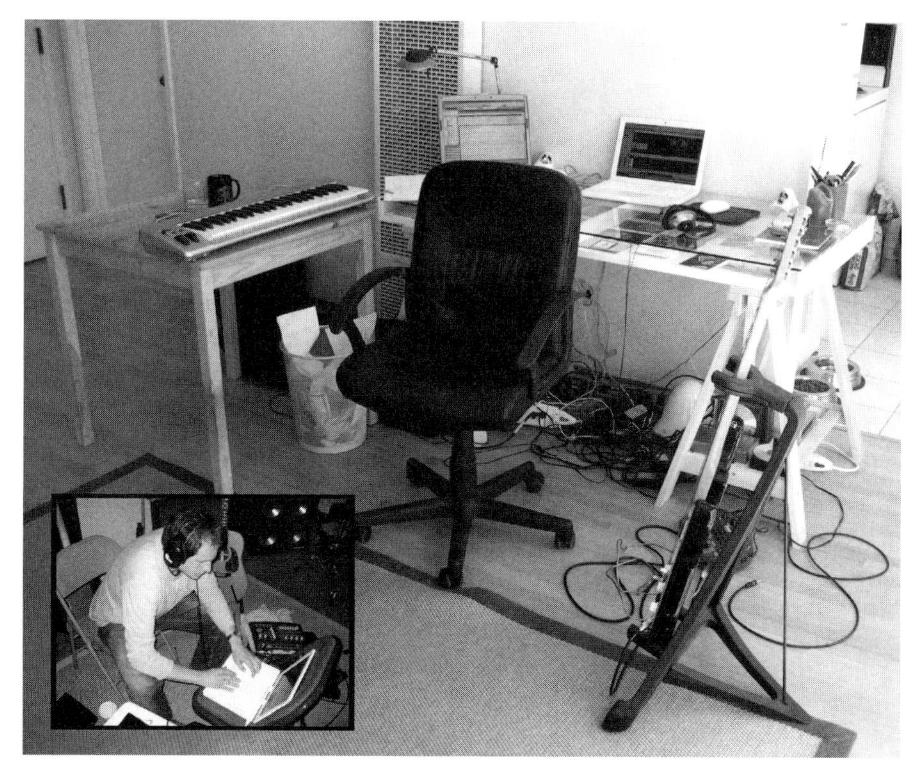

FIGURE 4.1
The author's previous (and very modest) digital audio workstation (recording remotely, inset). What's important for beginners to remember is that it's not how much you spend on gear, but rather knowing how to get the most out of what you have. When you start out it's best to keep it as light, portable, and inexpensive as you can. (Photo by Peter Dominguez).

A project or home studio

A project or home studio is run by a serious hobbyist or semi-professional. There are a lot of talented people in the world who simply can't afford or choose not to record bands full-time—it's a high stress, low-income profession at most levels. When you find a good, experienced enthusiast or semi-pro, hold on and never let go. These types tend to know a lot about their extensive stockpile of gear and the recording process in general. Most are musicians themselves. They may not always be the best arrangers or musically brilliant minds (and here's where an accomplished *producer* can add value), but many are very capable *engineers* (those who specialize in the technical side of the recording process) and can help you put it all together during the recording process—even if the final product needs refinement with input from you, your band, and/or an outside party. Be sure to request some of their recordings before committing, because some of these types are just tinkering gearheads with expensive mic collections and couldn't make a decent record if they tried. The good ones, though, are a treasure trove of audio and production knowledge and truly love what they do—and they do it well (Figure 4.2).

FIGURE 4.2 Musician/engineer Ron Guensche at his project studio in Northern California, New Future Vintage (Photo: R. Turgeon).

Pros: They do it for the love, not the money—in fact, you might even be auditioning for *them*. Homely atmosphere and a typically low $20–30 rate = a fun, penny-wise and comfortable experience. They also have the time, patience, and (oftentimes) inclination to field your questions about what they're doing and why they're doing it that way.

This set of advantages pertains to our next "pro studio" category as well, but if production is *not* your thing (and for the majority of musicians, it's not), having a pro record you is potentially much faster and easier than trying to produce yourself. Recording with someone else at the helm can be more fun and allow you to focus on coming in and nailing your parts—which you should be prepared to do in any outside studio (i.e., not yours) situation. In short, your project studio engineer can provide higher-quality hardware and the technical experience and expertise so you work fast and can stay creative.

Cons: Although project studio owners are often musicians, they may not excel at arrangements or making a record sound "complete," or slickly produced. Another potential issue is that of scheduling: most project studio owners have day jobs and have limited time to book around the band's schedule. Additionally, recording usually comes in "short bursts" after hours, in which case the clock is ticking at an even faster rate than it would feel like at a dedicated studio with ample staff and time for longer sessions.

A professional studio

There are different tiers of pro studios, but a professional studio is any facility—home or site-based—where the owner/operators record music full-time. These facilities produce records that go on the radio for indie and/or one of the 3–6 mega major labels (less likely with the full-time "home" studio); they've at least recorded radio-ready releases for top local bands. The full-time studio will usually have one owner/operator-producer and maybe an assistant/second engineer. Bigger studios will be staffed with full-time engineers, producers, assistants, maybe an intern or two to make coffee runs.

If your band is well practiced, you maintain your aesthetic vision, stay on budget and have a good engineer, you should be able to walk out of a pro studio with a quality demo. You're paying a higher hourly rate because, ideally at least, this facility and its personnel are more experienced and have proven track records. While it's entirely within reach to eventually make a great-sounding production on a limited budget (and one of the

FIGURE 4.3
Critically acclaimed San Francisco indie rocker John Vanderslice started Tiny Telephone in 1997 to provide affordable hi-fi recording to San Francisco's independent music community; the studio has a policy of setting recording rates under market price (Photo: R. Turgeon).

main purposes of this book is to encourage that notion), choosing a pro studio is a good way to go for less experienced recordists, provided you do your homework first (Figure 4.3).

Pros: Your demo will sound professional. The studio personnel may take an interest in your work, opening up a valuable contact/relationship for you. You'll learn a lot just by watching professionals work and benefit from their advice and coaching throughout the experience. Just keep in mind that regardless of how well regarded the studio may be, you should still do your homework: interview the staff and ask for demos. You probably don't want to spend hundreds or thousands of dollars at a studio known for its hard rock production style if you're music requires a more delicate treatment.

Cons: With hourly rates between $50–100 or higher per hour, pro studios can be quite expensive. Unless you don't need to worry about the studio bill for whatever reason, working against an expensive clock can create an atmosphere of pressure for everyone involved. While creative and intra-band tension is often a good thing, matters of money can often wreck bands and friendships—the two often going hand in hand. If you go the pro studio route, it's even more important for you to make sure that everyone's pocketbooks and goals are aligned before going in.

Is surround sound the next big thing in home audio and recording?
(Short answer: Maybe, but probably not and definitely not anytime soon.)
Six-channel (and greater) surround sound allows producers and engineers to mix a production to sound like playback is "surrounding" you from all angles. This makes for a neat experience, but it still seems to be for a niche audience now and for the foreseeable future. The average listener doesn't seem to need the supposedly "richer" audio experience surround can give you. Even though the prices of multi-channel systems have dropped significantly, two-channel stereo systems still seem to be enough. Also, because there's a limited demand for music mixed in surround, there's a limited catalog to choose from, and surround is simply more expensive for the labels to record and produce. Not insignificantly, the portable listening experience—whether it be an iPod or an iPod-playing boom-box—is a two-channel operation: People use their headphones or want one thing to carry like a boom-box when they're listening on the go.

The verdict: Because six speakers are a more complicated—if not a more expensive—way to listen, it's not likely that surround will catch fire in the next several years. Don't get caught up in the hype: learn to record and produce in two-channel first, then worry about the tsunami of demand for surround the marketers would like to happen. And learning to produce in two-channel sound is still the best way to learn and master the fundamentals.

LOCATION

One thing to keep in mind is that audio production is very portable these days, and as great producers like Rick Rubin and Daniel Lanois (U2, Peter Gabriel) know, location can make a big difference in the final result. If you or the engineer/producer you hire can work remotely, think about recording in a location that inspires creativity while being conducive to good acoustics. Rubin's work with the Chili Peppers in his Laurel Canyon mansion on their breakthrough record *Blood Sugar Sex Majik* redefined their sound and made them household names, the mansion still being the place that lead singer Anthony Kiedis considers "the address of his job." Even if you don't get to record in a cool mansion with a rich rock heritage and deep, mystical vibe, at least pick a place that makes you comfortable.

Along with location, it's time to consider one more thing before we get into the nuts 'n' bolts of the recording process (Figure 4.4).

FIGURE 4.4
Boston's Karate (left to right: Geoff Farina, Gavin McCarthy, Jeff Goddard) recorded several sessions in a living room so cozy it helped inspire the "In My Living Room" compilation on which Karate appears (www.kimcheerecords.com/catalog/kc004). (Photo: Andy Hong).

THE ROLE OF THE PRODUCER

If and when you ever get signed to a major label, the label will want to hire a pro producer (qualified below, see "Warning") to supervise the recording of your record. If you're planning on recording yourself, working with a project studio or a pro studio, you may choose to be the producer.

Serious musicians and music fans have very strong opinions about producers because they know how much influence a top producer can have over a record and even a band's overall direction and sound. Definitions and the influence and role(s) of today's indie/mainstream rock producer vary, but it's conventionally analogous to a film director: they are not necessarily "the boss," but they are the ones that bands—and more often—record labels hire to make a record a success. They are typically involved in every stage of the record production process, from making sure the recording stays on time and budget to choosing and arranging a band's material to recording, mixing, and mastering. Where a recording engineer conventionally works without a lot of creative (but more technical) input at the service of the producer, it's fair to define today's producer as the "non-band member" most responsible for shaping a record's overall sound.

If you want to understand the influence a top-caliber producer can have on a record (along with a top-caliber studio environment, gear, and production team, of course), withhold any subjective opinions and listen to Scott Litt's lush production of Juliana Hatfield's *Become What You Are* and compare to a more raw and spare production like *Bed*. Listen to Nile Rodgers' rather big-sounding production of David Bowie's *Let's Dance,* on which he also played guitar, then any other Bowie records. Listen to Green Day's first two records on Lookout Records, then their major label debut, *Dookie,* produced by studio veteran Rob Cavallo. Listen to Weezer's jubilant but raw, sloppy, and unfocused *Pinkerton* before enjoying the almost sanitized-sounding, pristine, and polished pop of their first and third records produced by Ric Ocasek (Weezer's third "Green" album being even more finely-tuned than their first).

Warning: Beware of the self-proclaimed producer
"Producer" is one of those titles that's more self-conferred in the film and music worlds than we'd like it to be. Just because a guy can make a beat or two on a drum machine or knows a bit about Pro Tools doesn't qualify him as a producer. If he or she doesn't have a demo and track record, engage with extreme caution—especially if your money is at stake. And if he's getting in the way at your sessions in terms of not contributing—or contributing in all the wrong ways (i.e., useless or unwanted input)—can him immediately.

Litt, Bob Rock (Metallica), Rick Rubin (Run DMC, Red Hot Chili Peppers, Johnny Cash, Weezer), Mutt Lange (AC/DC, Foreigner, Def Leppard), Daniel Lanois (U2, Peter Gabriel) are all producers who have worked on some of these bands' best known records. It's not unusual to hear a highly successful band gratefully acknowledge their highly successful producers as the "fifth" (or sixth, etc.) member of the band, acting behind the scenes as the one making a lot of critical aesthetic and technical decisions that can—as you've heard—make all the difference. Some producers like Rock literally do double-duty as band members, or more commonly, perform musical parts on recordings.

To understand and recognize the importance of the producer's role, try to identify records not only by band, but by those producers you admire. See if the records you already like have a "common denominator" of a certain producer or producers. Find articles and interviews with the producers of the records you like the most.

Even if you decide to make records with a producer other than yourself, it's highly beneficial to understand production fundamentals and vocabulary to gain better control over your sound. Regardless of whether you're recording your own band or another band you like and admire, you may find that production is a good avenue for you in place of or in addition to writing and performing your own music. Other big advantages of taking a more "behind the scenes" production route to your music career can include the (relatively) regular hours and stationary location (i.e., no touring). Plus, you get the variety of recording different acts in different genres, if that's important to you. And after hearing how much of a difference a professional producer can make, you'll realize just how creative audio production—a truly rewarding and fascinating blend of art and science—can be.

Now that we've covered pre-recording considerations and have a basic understanding of studio options and producer roles, let's talk about…

CHAPTER 5
Gear

DON'T OVERDO IT

Let's assume you've decided to take the D.I.Y. route and make your own record. Before we start itemizing the basics you'll need to record, it's worth reemphasizing a point we made in the beginning: with a little talent, knowledge, patience, and tenacity—and a very modest amount of gear—you can make a record that sounds quite good.

The bottom line is that if you're just starting out, it's highly recommended that you *keep your overhead low*. Much of today's recording technology is extremely powerful and, in the right hands, can deliver results that rival the biggest pro studios—and for much less money than was ever possible.

That said, be wary of magazines telling you what "essential" gear you need. Instead, get what's right for your skill level, preferences, production needs, and budget at any given time. Keep your focus on being creative and producing—not on trading, trying, and buying every new piece of gear.

> **Where to buy**
> Although you can buy gear at any music store, at pawn shops, or online, it's worth noting that it can be helpful for musicians at any level to talk to *friends*, check out *message boards* (like Tape Op's at www.tapeop.com (>"Message Board"), tweakheadz.com, and gearslutz.com. Also search "pro audio forums.") and read the *magazines* (see the Appendix A, Audio Production Magazines, at the end).

FIGURE 5.1
Seattle's Spanish for 100 browsing on tour at Midwest Buy & Sell, Chicago (Photo by Ryan Schierling).

HOOKING UP: THE SIGNAL PATH

Since this book largely assumes you can teach yourself your hardware and software with the included (or third-party) manuals, we're going to stay general here and not get too specific. There are of course an infinite number of ways to put together your DAW based on your preferences and hardware, so Figure 5.2 shows what most beginning home recordists start out with, along with a few common extras.

One thing all these setups have in common, though, is a signal path—the path through which the sound travels from the source onto a recording medium (which would be a hard drive (HD) in the digital domain and magnetic and reel-to-reel tape in the analog domain).

Every recording requires:

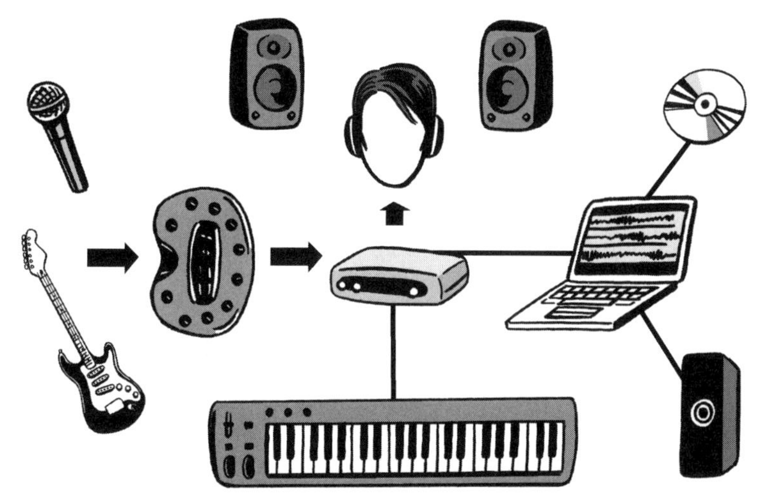

FIGURE 5.2
Here are the basics and how they hook up: You record a sound source like your voice or an electric instrument into an audio input device, possibly with a guitar modeler or effects in between. You listen via the audio input device's (or your computer's) stereo output via headphones and/or monitors. The input device translates the real world signal to digital and passes that information along to the computer, via a data line (USB, FireWire, PCI(e)) to an internal or external hard drive. If you have a sound source, a computer, headphones, and the audio input device, you're ready to record (Illustration: R. Turgeon).

A sound source—Real-world instruments

A sound source is made up of one or several of the following:

Real-world instruments like a singer, amplifier, or acoustic instrument (tambourine, acoustic guitar, piano, strings, brass, woodwinds, etc.); or any electric instrument recorded directly via their output ports (electric guitar, synthesizer); or soft synths: instruments that exist as software, and are typically controlled via MIDI. Both types require the proper cables running into the input units of…

1. A *pickup device* like a microphone or line output
2. *An input to a recording device* like an audio input device (covered below) and

3. *The recording medium:* This is what the sound is actually recorded onto, like an **HD** (digital) or an **analog** recording medium like reel-to-reel tape.

GEAR LIST AND HOW IT WORKS TOGETHER

Here we'll cover the gear that constitutes most digital audio workstations, from beginner to pro.

Instruments

These would be the real-world or electric instruments described above. For versatility and variety, you can't beat the digital domain, since MIDI (more on MIDI ahead) lets you convincingly emulate almost any instrument (except maybe vocals and guitars). Even guitar amps can be reliably emulated by *amp modeling* software and/or hardware like the Line 6 Pod (see Appendix C, Recording Process and Gear List for more on the Pod), so you can record extremely convincing guitar parts without disturbing the neighbors.

A computer

We are past the tipping point: CD sales continue to plummet. Big studios and record companies are hemorrhaging money or have bled dry. (Good) bands are achieving global exposure online. The music world—both musician and listener—live and thrive in the digital domain. To produce and promote one's band, today's independent musician really does need a computer.

Fortunately, powerful, portable computers are relatively cheap these days: you can buy a decent Mac or PC laptop for under $1,000. Do your research, but most low-end laptops have enough processing power for multitrack recording, at least enough for a rock record. Get as much RAM and processing speed as you can afford—doing so will help postpone the eventual "march-of-technology" upgrade down the road. Knowing how to produce your own records is a great investment in itself, but the computer will be an invaluable tool towards helping you promote your music online; listen to and download new music; network with bookers and bands; learn about new opportunities; research and buy new gear; and do just about everything else related to recruiting, recording, and promoting your band.

Something else to consider is getting a computer that's largely, if not solely, dedicated to your audio production endeavors. It's not *necessary*, mind you, but the more software you have on your computer, the more it potentially interferes with or slows down your audio operations. Don't fret over this at all until you get more serious; it's just something to keep in mind after you start getting good...

Software

A lot of very good recording software comes free with most audio input devices, and it's a great way to get your feet wet. As you get more advanced, you'll probably want to look into more feature-laden, "pro"-oriented packages like Pro Tools, Cubase, Sonar, Live, Logic, and many others. Research audio *plug-ins* (more on these is discussed in "Recording," Chapter 6) online for some great effects and mastering tools. There is no right or wrong software to use, only what's right for you. Most software packages have fully functional demos that you can download online, making it easy to try before you buy. They also tend to be "tiered," meaning they scale up in features and power with correlating product identifiers (Pro Tools, Cubase, and Logic all have "scaled" versions, and some "light" versions come bundled with audio interfaces). Lastly, instead of burdening yourself with learning every new software package that comes out, focus on learning and mastering one. While there's nothing wrong with eventually building a wider range of audio tools to work with, you'll work faster, smarter, and get better results when you start to master just a few programs and plug-ins.

Audio input device

This takes in your sound sources and converts their analog signals to digital for recording onto your HD. Smaller audio input devices tend to have fewer inputs (typically 1–4) and are more portable. Larger input devices typically have more inputs and can be stand-alone or rack-mounted. They typically plug into your computer via a *USB* or *FireWire* connection. While neither has inherently better sound quality, Firewire devices tend to be more stable than USB in general. Some feature mixing boards, extra headphone outputs, and other bells and whistles.

Some considerations here include number of inputs that can be recorded simultaneously, mic *pre-amps* (does it have them, and how many), number of headphone outputs, built-in instrument modeling, and MIDI i/o (input/output). With the first two items, quantity and price typically correlate. Unless you're recording drums or full bands, you generally won't need more than a pre-amp or two at a time. If you plan to record drums or full bands (either in the studio or "on location," like at a show), you'll need a more expensive model with more inputs and preamps. If you don't need to worry about these situations and plan to record drums or other multiple-input situations at a project or professional studio, a box with one mic and/or guitar input should suffice; these 1- or 2-input units (usually 1 XLR and ¼" for mic and guitars, respectively) are a lot less expensive and much more portable than their multiinput big brothers. One feature that will make your life easier if you plan on tracking vocals is phantom power; you'll likely want a condenser mic for vocals, and the vast majority of them require phantom power to operate.

Hard Drive(s)

It's always best to record audio on a HD that's separate from the operating system (OS) if possible, whether it's internal (inside the computer) or external (one that plugs into the computer via a FireWire or USB connection). This helps to minimize the potential for interruptions of audio data being recorded and played back as virtual memory or background tasks related to the computer's operating system access the OS HD.

Another compelling reason to get not only one but two external HDs is that you should back up your files pretty much every time you work on your music; so you'd ideally record on external HD 1 and backup to external HD 2. Now while you can record your tracks using your computer and no external drives at all, saving less than $200 or so on at least one external drive is not really worth losing the creative inspiration and time you may have invested in work lost to accidental deletion or malfunction; it's not a corner worth cutting if you can help it. One last thing to consider is that most external HDs come with free backup software that makes backing up a session's work fast and easy, and if not, there's plenty of backup freeware online (MacUpdate and Downloads.com are two good sites for low-cost or freeware). In lieu of an external drive or two, it's worth mentioning that burnable CDs (somewhat impractical, limited storage) or DVDs (up to about 5–6 MB of storage, and DVD—RWs are rewritable) also provide a reliable, durable backup, and data transfer medium when you're working on more than one computer.

Some considerations that affect HD prices are speed (get at least 7200 RPM for audio production, along with an 8–16 MB buffer), portability (smaller = more expensive), the type of interface (USB, FireWire, or both), overall durability (you'll usually pay more for "high-end" design HDs that look like fireproof little tanks) and, of course, the amount of storage itself, expressed in gigabytes (e.g., 150 GB).

A typical 10-song rock record—including all takes and final tracks—usually takes up 10–20 gigs of space depending on your production style, file maintenance habits, and number of instruments recorded. So while more gigs are always preferable, a drive that's anywhere from 80–160 gigs should be more than enough as you get started. Some better known HD companies are LaCie, Maxtor, and Seagate, but there are plenty of manufacturers out there to choose from. Again, these are nothing to agonize over and you'll be fine with, say, a HD with specs like: 150 GB, FireWire, 7200 RPM, 8 MB buffer. 10K RPM is overkill for audio production, and due to louder operational characteristics, isn't recommended.

External CD—DVD burner

Even lower-end laptops include CD burners these days, with more expensive computers housing a "combo drive," meaning it burns CDs and it reads/plays back DVDs. Even so, consider an external burner because you will likely be burning lots of demo CDs to pass out to fans and bookers, and doing so often will create a lot of wear and tear on any burner your computer might have. It's better to reserve your computer's CD/DVD slot for lighter operations like software installations and audio or DVD playback, not extensive CD burning. If your computer's CD slot goes bad, it can be very expensive to repair, not to mention incredibly inconvenient in terms of time lost while waiting for that to happen. CDs burned from laptops' internal drives also seem to face more compatibility issues than those burned using dedicated, external burners.

Cables

If you have a project studio, you may only need to buy as few as two: a quarter-inch instrument cable and a microphone cable.

There are two types of cables: analog and digital. The digital cables you'll be using will largely be of the USB or FireWire variety, and for the most part, the ones that "come in the box" with your digital gear will do just fine. Why? Because data is just ones and zeroes—not a more complex analog signal—and the "quality" of those connections generally doesn't affect the "quality" of your data transfer, let alone your recording.

Mic and instrument cables, however, are an "analog" purchase—meaning they create or transfer an analog audio signal, not digital data—so potentially they have much more positive or negative impact on the analog vocal or instrument you're recording. For the most part, any mic or guitar cable will do as long as they don't create audible buzzing, humming, and noise. However, analog cables are like other "analog purchases" like guitars, drums, or amps: you get what you pay for—to a degree. A $2 cable from the local music or electronics store can ruin your recording if it doesn't have quality connectors, adequate soldering, or proper shielding (this keeps airborne interference out of your signal chain). Most cables with metal ends are going to be acceptable, whereas most cables with molded plastic ends are junk. If a cable is made with shoddy materials and badly constructed, you may hear a lot of noise or a weak bass signal; this is most common with cables with plastic ends. Like every other component in the signal chain, quality is important, but be very careful not to overspend unnecessarily.

Monitors

Monitors are essentially speakers designed for the specific purpose of listening to audio playback during production. Since they'll give you the most accurate reproduction of sound without coloring it the way general consumer speakers can, it's recommended that you use them to mix and master. There are as many monitors out there as there are stars in the sky (believe it or not, it's not the niche market you might think), the most commonly known types being *near-field, mid-field,* and *far-field,* with "active" (or powered) and "passive" versions of each. Active monitors have an internal power amplifier, whereas passive monitors require an external amplifier.

Near-field monitors are designed to be listened from just a few feet away (i.e., where the influence of the room's walls on the sound is minimized), so they're perfect for use on desktops and in tight spaces. Mid- and far-field—both active and passive—are usually larger, more expensive, and used for louder, more "general" playback. Since active near-fields also have their own power supply, they're more "self-contained" and portable, making them the natural category to choose from for home studio recordists. You'll of course want to primarily consider any monitors' "footprint" (i.e., size in relation to your available space and setup) before how well they work for you from an audio standpoint. Regarding where to place your monitors: A simple rule of thumb for near-fields is place them at about ear level, maybe 2–4 feet away, equidistant from your ears and from each other.

You can get a usable pair of near-fields for around $200, but before you commit to any master, you will definitely want to burn a CD and play your recording through different systems: headphones, boomboxes, car stereos, etc., and take notes. The goal with final mixes and masters is to get them to sound as good as possible across many different systems. This is where mixing and mastering on monitors—and listening on other speakers for fine-tuning—comes in handy.

One final consideration in this area is the use of a *subwoofer* in the production environment. If you use one, keep it on the floor and at a distance equal to the near-fields or where the manufacturer recommends placing it in their product literature.

Mics/pop guard/mic stand

Mic collections in studios can include dozens of brands and types and be worth thousands of dollars. If you're just starting out, don't plan on recording drums (more on this later), and don't want to spend a ton of money, you may only need one or two basic-level mics to record vocals, guitar amps, and other real-world instruments.

There are several types of mics, but the two most commonly used are *dynamic* and *condenser.* Dynamic mics are often cheaper, more durable, and don't require a power source, which is why you often find them used in live settings. Condenser mics are often more fragile and require a power source. Most singers prefer their vocal tone through a good condenser microphone, but a dynamic is fine if you're just starting out. Don't fret if you can only afford a (dynamic) Shure SM57 or SM58 for both live and recording situations—they'll do the job, and rock musicians around the world have proven that SM58s especially can withstand a good deal of abuse. Incidentally, every mic type has a certain directionality or polarity pattern—*cardioid, omnidirectional*, and *bi-directional* or *figure-8* (see Figure 5.3) to name a few—that denotes the area or soundfield that the mic is sensitive to or "picks up." Cardioid mics pick up mainly what's just in front of them, making them ideal choices for vocals, guitar amps, and drums. Omnidirectional mics pick up sound in all direction, so they're typically used for groups of instruments or drum overheads. Bidirectional or "figure-8" mics capture sound from both the front and back while rejecting the sides, so they're typically used to pick up two instruments in close proximity. Pending further exploration on your own, you really only need to be concerned with a good cardioid type for close-range recording on something like vocals or a guitar amp. As you start recording, you'll find that while the fundamentals aren't too tough to grasp, the mic's placement can make an enormous difference; it's certainly something you'll want to experiment with as you record (...and some people spend their whole lives doing just that. More on mic placement for specific instruments ahead in "Recording," Chapter 6).

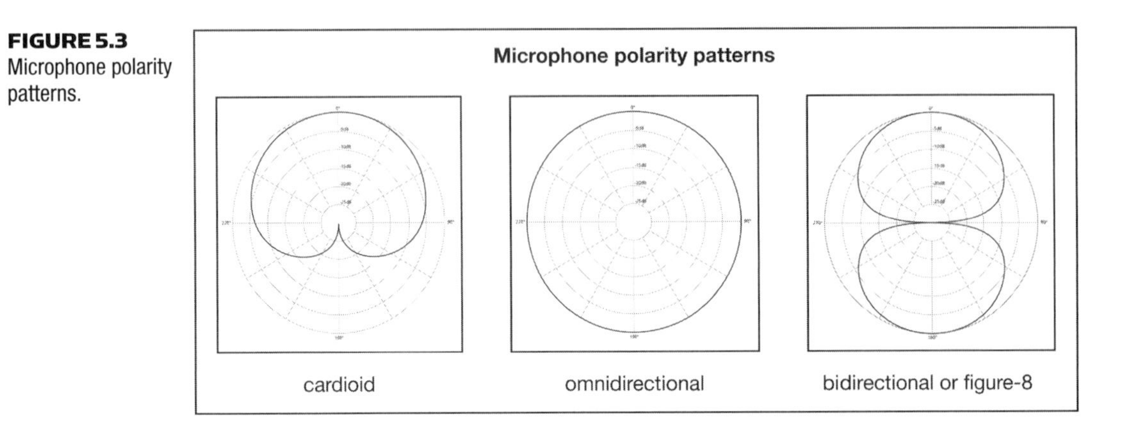

FIGURE 5.3
Microphone polarity patterns.

As far as recording vocals go, you'll need a mic stand and a *pop-guard* (a small nylon or metal screen you clip to your mic stand for placement about 2–3″ in front of your mic) to help control the breathy "pops" of your 'P's and 'B's (plosives), which are very hard to remove after being recorded and can render the track unusable (more on the de-esser plug-in, which can help remove the harsh "S" sounds from vocal track, ahead in the Mix section). (Tip: Pawn shops are a great place to find mic stands as it doesn't matter if they're used). When you record vocals, clip the pop guard to your stand and place it a few inches in front of the mic like in Figure 5.4 and you're all set.

FIGURE 5.4
Mind your 'P's and 'Q's: The almighty pop guard (Photo by R. Turgeon).

What mics should I start with?
As we said above, people just starting to get into recording may only need to start with one or two good mics. Arguably, the most ubiquitous mics are the Shure SM58 and SM57, both dynamic/cardioid mics that are known for their solid value and reasonable price. The SM58 is typically used for vocals, and the SM57 for acoustic instrument and amp micing.

If you're willing to spend more on mics, you'll probably want to invest in a good condenser mic to cover lead vocals and more delicate sounds and acoustic instruments like guitar and pianos. You can start by checking out the Apex 460 ($230), Audix CX112 ($400), Audio-Technica 4033 or 4050 ($200–$500 used/new), and the Rode NT1a ($230). Some of the best known mic brands you can also start checking out include AKG, Sennheiser, Neumann, and M-Audio.

MIDI keyboard

MIDI stands for *Musical Instrument Digital Interface*. A MIDI device is most commonly a keyboard or synthesizer that plugs right into your computer, and you can buy it for as little as $100. Describing the possibilities and technical operation of MIDI devices, software and hardware can get quite technical, but you can think of MIDI as a common language in which compatible hardware and software communicates. What's important to note is that MIDI itself is not an audio signal—it's a digital signal that contains note information. To hear MIDI, either a real or virtual sound module must convert incoming MIDI data to an audio signal.

To illustrate with a practical example, you can record notes played on your MIDI keyboard onto your computer, then manipulate them in any number of ways:

- You can pretty much *emulate any sound* you want—trumpet, flute, big 1980s synth, harp, piano—you name it, then, if you like, switch it to any other instantly.
- Some software will *instantly score your recorded music*, which you can then modify on-screen in real-time.
- You can instantly *modify the musical characteristics* of MIDI notes themselves, adjusting all or some of their duration; *velocity* (i.e., how "hard" each note is "hit"); reverse their pitch and position; *humanize* (randomize them a bit for more "feel" so they sound less rigid); *quantize* notes on the tempo grid to get timing absolutely precise; make them crescendo or decrescendo; and a whole lot more.

If you want to have sounds and virtual "instruments" above and beyond real-world, organic vocals, guitar, drums, and the like, a MIDI keyboard is a cheap and valuable investment. If you're not a keyboard player per se, though, it should be noted that you can just as easily use your computer's keypad to record simple notes (or manually place them with your mouse) for similar manipulation with software as basic as GarageBand.

Headphones

You will certainly use headphones to record in any studio (it's usually how you hear background tracks when overdubbing), but if you start producing yourself in a home studio where you have to mix without disturbing the neighbors, you will begin to appreciate headphones in a whole new way. Studio headphones (vs. consumer headphones) are like monitor speakers in that they're designed for the most accurate and least sonically "colored" playback as possible, and there are a variety of types and models designed for specific studio applications. Some are meant to be as accurate as possible, while others are meant to provide maximum isolation, "isolating" either instrument noise from the outside "leaking", in or headphone playback from leaking "out" onto the recording. If you're a drummer, you'll definitely want to check out isolation headphones like Vic Firth or Direct Sound Extreme isolation headphones so you can hear any input—whether it's live, playback, or both—and not the din of your drums bleeding through.

For general purpose studio use, it's a good idea to find *closed-back* headphones with a neutral response. *Open-back* headphones A) "leak" sound from the headphones, which you don't want to pick up in your recordings, and B) don't isolate the outside world as well, which can be distracting if you're mixing. You could mix and record with iPod earbuds, but doing so will compromise your mixes since you won't hear the fuller, more accurate sound better headphones provide.

Using inferior headphones to mix can often result in mixes that are inaccurate when played back on speakers. For example, people tend to mix in more reverb when using headphones than they would with monitors because, acoustically speaking, the room isn't being taken into consideration. The result is that when the headphones mix is played back on speakers, the mix sounds too wet (see sidebar below).

Even if you invest $30 in a better pair of headphones with which to record and mix, it's worth it. A good pair of studio-grade headphones will cost around $100 or more. But don't just go by price—read reviews online and absolutely listen to them yourself before buying.

Why you shouldn't only mix on headphones

Sound travels through air, and the character of that sound is largely determined by external physical conditions: the room you're in, the hardness of the walls, ceiling, and floor, etc. This is why music sounds different when you hear it on speakers or when it's wafting through the window from the flat next door. Even the same speakers and sound source can sound very different when placed in a different room, let alone when the source is played through different speakers or headphones. This is why professional studios and mastering facilities place as much or greater emphasis on the listening environment as they do on their gear (room dimensions and materials, speaker type, and placement); their rooms are often custom-built to achieve as much sonic "accuracy" as possible. Pros know the distinct and unique "reference" sound of their room and monitors, and they have to.

So while rough-mixing on headphones for a stretch is fine when you don't want to disturb the neighbors, be sure to finish your mixes on monitors in your own "reference" environment. Hearing music played back in a real physical space through monitors (as opposed to only hearing it through headphones) is critical when you're working toward getting a mix that sounds best across the widest variety of systems, speakers, and headphones included. You'll also want to be sure to listen to your mixes on different systems outside your listening area: your car stereo, a boom box or two, and different monitors, to cite some examples. When you get them to sound good across a variety of speakers, you've nailed it.

CHAPTER 6
Recording

Before we get into the particulars of recording individual instruments and sounds, let's start off with a few basic recording concepts you should become familiar with.

HEADROOM

Headroom is a term referring to the amount of "room" you have between your audio signal's maximum volume and 0 dBfs in any given channel (dBfs is deciBels full-scale—a measure of level in the digital domain where zero is the top of the scale and all level values below that are referenced as negative numbers). Even though you set your hardware or software faders to control gain and relative volume for each track, the "maximum" volume for individual channels and the overall mix is of course typically dynamic, changing with each kick drum hit, guitar note struck, etc., and between silent and loud passages. Any channel is *peaking* when it hits a high or the highest level during playback (a *peak* being a high or the highest point in any audio waveform).

You control your headroom by (A) adjusting the input gain of your input signal and (B) adjusting your channel or master volume. When you record, you'll want to make sure you're not red-lining (i.e., going above 0 dBfs), which typically results in a harsh popping or crackling sound and unusable tracks. When you record or mix, you'll want to leave yourself about 12–18 dBfs of headroom so you can make adjustments without peaking in any specific channels across your mix and especially in the final mix. For final mixes, you'll want to leave about 3–6 dBfs of room on the master bus to give the mastering engineer (or you) enough headroom to play with compression, limiting, and other settings that will help bring your mix home.

RECORDING WET VERSUS DRY

A single track or, less commonly, master track, that's described as being "wet" typically refers to the amount of reverb or overall "ambience" (combined effects like reverb, chorus, and delay in particular) applied to it.

You can make tracks "wet" after the fact in the mixing stage (more common), or you can *record* wet (less common), where you'd route your input signal through your effects as they're being committed to your recording medium. You can also route your input signal through preamps, EQs, and/or compressors that can amplify and color the sound before it's recorded.

There are pros and cons to recording wet or dry, all dependent on the situation and sound you're going for.

Recording dry gives you the flexibility of being able to apply and change effects in the form of plug-ins in your software in the mix stage after you've recorded.

Recording wet can save time in the mix stage if you want to apply effects (or coloration by routing through compressors, EQs, or preamps) to the signal before it's recorded. The disadvantage here is that if you don't like how it turned out or it's not working in your mix, you typically have to re-record the track(s) because it can be difficult or impossible to reshape a wet signal into what you want.

It's recommended that newer recordists err on the side of recording more dry than wet—you can always color the sound more to your liking using software plug-ins in the mixing stage. Even experienced recordists think twice before recording wet, or with EQ or a lot of compression applied to the input signal since it's nearly impossible to restore the original, dry sound source's inherent sonic character. If you really want to record wet or with EQ or compression going in, you can always record the dry signal on one track while routing the wet signal to a separate track.

A related point to keep in mind has to do with *recording reality* versus *emulating reality*. Although the digital domain gives us a virtual symphony of sounds, amp modelers, and effects to help us apply the perfect color or character to MIDI data or any real-world sounds, there's much to be said for keeping it simple. If you want to record a real-world sound like a beat-up piano, your Les Paul played through a Marshall stack, or your pawn shop bass drum, that's exactly what you should record. You can spend a lot of time (and processing power) looking for the perfect digital tweaks to your real-world sounds, but more often than not, you're better off studying mics, mic placement techniques, writing great songs, and working on your instrument(s)—and recording reality—versus relying too much on your PC to make your records stand out.

MIC TYPES AND PLACEMENT

As we said in the "Hooking Up" section of Chapter 5, "Gear", people can spend an inordinate amount of time and money in search of the perfect mic for any given situation, along with how their mics are placed with regard to the sound source. The important thing is to realize that as long as you have at least one durable, reliable dynamic mic like the trusty Shure SM57 or SM58, you can pretty much record any real-world sound, including vocals, amps, drums, piano, acoustic guitar, etc. (And again, if you want to get more detail and depth from your vocals and acoustic instruments, an additional condenser mic is a good way to spend your money even if you're just starting out.) You only *need* multiple mics if you want to put them on a single instrument like piano or drums, or you want to record a live band.

Mic type and placement can make a big difference in the overall tone, clarity, and character of the recorded sound. Even the most experienced producers and engineers never stop tweaking, experimenting, and learning more about mics and mic techniques to achieve their perfect mix. People prefer any number of mic types, brands, and placement techniques, so you'll definitely want to research and experiment on your own.

That said, what follows are some recording basics for individual instruments, including conventional mic type, brand, and placement. (This of course sometimes relates to how these instruments typically sit in the frequency spectrum and/or mix, which is why the "Mixing" chapter follows).

Drums

Before recording, tune your drums and make sure there's no ringing or rattling as it's very hard to eliminate that noise in the mixing stage. (If you don't know how to tune your drums, find or hire someone who does.) Next, it's safe to say that two typical drum mic placement patterns are:

- *3–4 mics total*: 1–2 overheads (i.e., a few feet "over" the highest cymbal or cymbals), one kick (or "bass") drum mic, one on the snare
- *6–7 mics*: Same setup as above, adding individual mics on the toms

There are practically infinite variations, combinations, and preferences around micing drums, but it's worth adding that some prefer adding condenser mics dedicated to the hi-hat and ride cymbal for added emphasis even though the snare and right overhead mics, respectively, pick these up.

Drum mics are available in kits like the Red5 Audio RVK7, but Table 6.1 lists the commonly used individual mics, along with conventional mic type and placement. Figure 6.1 shows a typical mic setup for a full kit.

Table 6.1 Drum mic reference

	Mic type	Placement	Commonly used mics to explore
Snare	Dynamic, cardioid	Almost touching the snare head or a bit higher depending on your preference—just make sure it's not catching too much hi-hat. You can also mic the bottom head.	Sennheiser E604, Shure SM57
Rack and floor toms	Dynamic or condenser Cardioid or hypercardioid	A few inches away to very close to touching the heads (often a matter of preference). For multiple toms, in between the toms a few inches (or closer) above the heads. You can also remove the bottom heads and mic the insides of the toms.	Kel HM-1, Sennheiser E604, Shure SM57
Kick drum	Specially-built dynamic	Most engineers I've worked with take the outside head off and dampen the kick with a pillow or blanket, then place 1 or 2 mics: ▪ a few inches away and off-center from the front head ▪ inside the drum, a few inches away from the inside head and a bit off-center from the beater	Audio-Technica AE2500, Audix d-6, AKG D112, Sennheiser E602, Shure Beta 52

Table 6.1 Drum mic reference—*cont'd*

Overheads	Condenser cardioid or omni	A few feet above the cymbals. Position these by ear to achieve a balance between the cymbals, ideally keeping the snare in the middle of the stereo image.	Audio-Technica 4033; AKG C1000, C2000, C451, C3000B, C414; Kel HM-1, Rode NT-4; Shure SM81

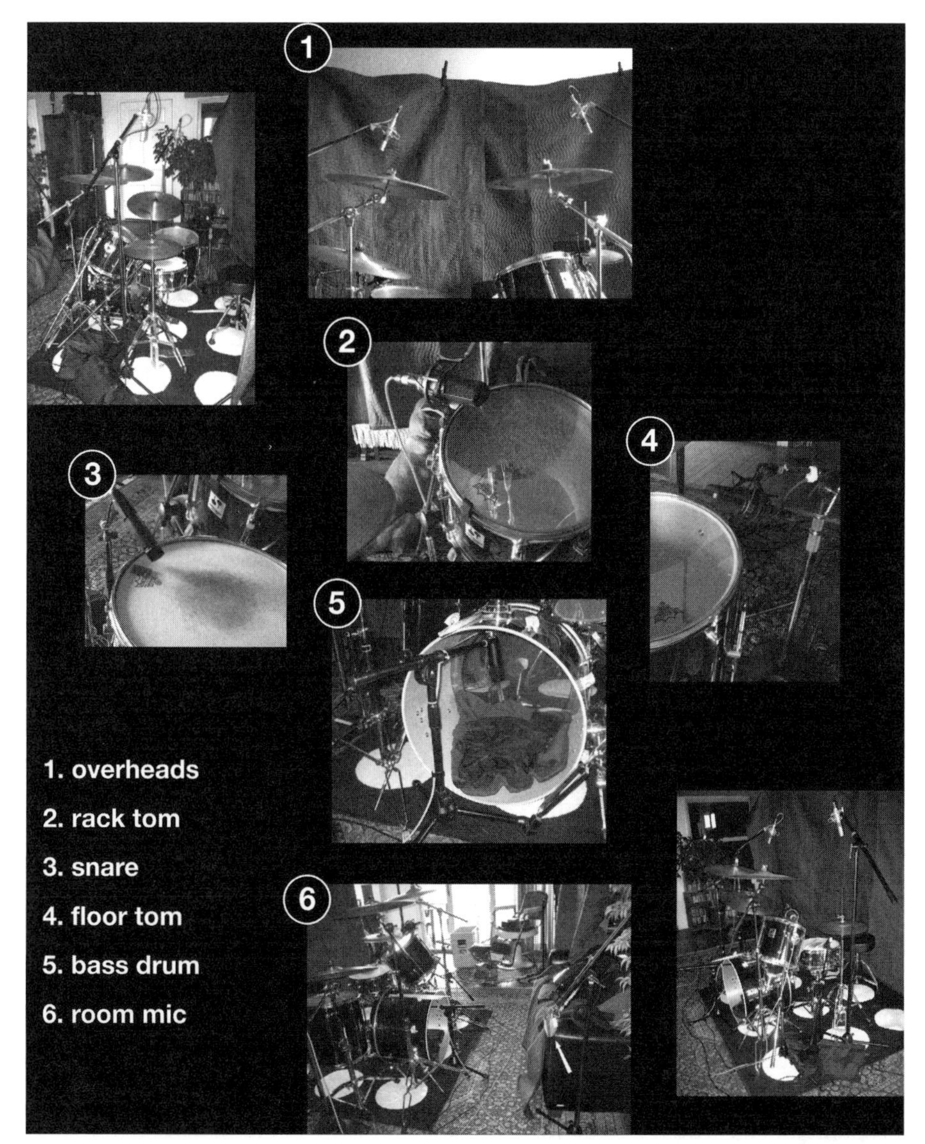

FIGURE 6.1
Drum mics (Photos/design: R. Turgeon).

To click or not to click

Should you use a click track or not?

Again, a subjective question, but if you want your recording to sound as polished and professional as possible, producers typically ask the drummer to use a click. A click set to a tempo and click track (expressed in a BPM number, or beats per minute) makes it easier to edit tracks and manage the overall meter or the song, which, in radio-friendly music especially, is largely about maintaining a steady tempo on the drummer's part. Steady tempos on every player's part simply make the final recording sound more professional and "finished" than it would without using a click. Another advantage to using a click is that when everyone's tracks are mapped to the "grid"—or the strict tempo along the software's set tempo—it makes it much easier to find, edit, and manipulate MIDI data. Finally, you can't apply modest amounts of *time correction* if there's no set "time" or "grid" to correct it to (more on this plug-in will be discussed in Chapter 7).

Some drummers (and producers and other musicians) will resist the idea of playing to a click track, usually claiming that it will interfere with their "feel" and the "groove" of the song. This resistance is usually due to the fact that the drummer isn't used to or doesn't have much experience with playing to a click. If you run into this, realize that playing to a click track does take some getting used to and that the key is giving your drummer (and/or band) the time and space to do that, making sure that they're comfortable with the volume levels of the click and any other tracks they're recording to. Once they get the hang of it, it's not difficult. It's also worth mentioning that just because there's a set tempo and click track, it doesn't mean the drum performance or overall recording needs to sound stiff; the click can just be used as a guide for the drummer to groove ahead of or behind.

Bottom line: If you're serious about working efficiently and wanting the best results from your recordings, a click is recommended.

Bass

Stick a sturdy dynamic mic a few inches away from the amp's speaker(s), or just record direct (straight from the bass through a cable to your audio input device), or do both. I prefer not recording a bass amp but instead going directly into the audio input device or soundcard (or "direct") because you can do it at home, and you can color the dry signal any number of ways in the software/mix stage. As with anything else, you also always have the option here of applying EQ or compression on the way in. Just remember from the beginning that what gets committed to your recording medium stays there, and adding effects to a dry signal in the mix stage might be a better way to go.

Guitar

Lots of options are here. You can:

- Record **a dry, direct signal** and apply plug-in (more on these ahead) effects, compression, EQ, and/or *amp modeler* software in the mix stage.
- Place a dynamic mic a few inches **from your amp**. This gives you the option of recording your real-world hardware pedals and effects you might have as part of your live setup. Even affordable Shure SM57 or SM58s are fine for recording a live guitar amp. Some people prefer to place a second condenser mic at the amp's edge, a few inches or feet away from the amp, or somewhere else in the room to record onto a second mono track as a "room" or "ambience" mic. Mic placement on amps can make a tremendous difference, so you'll want to experiment to get the sound you're after.
- Record direct using a Line 6 Pod or similar **amp modeler hardware** unit. This is my preferred method since it doesn't disturb the neighbors and you can record and re-record guitar parts to your heart's content at home. You also can use the Pod just for the amp modeling, without reverb and other effects going in, which gives you more room to experiment with software effects in the mix stage.

Production tip

Doubling guitars and vocals

A typical trick for thickening and sweetening guitars and vocals is to record the same part on two (or more) different mono tracks. We'll be covering how to mix these separate tracks for certain situations in the upcoming Mixing section, but for now, it's just important to register that laying down two of the same vocal and (usually rhythm) guitar tracks is a fairly common technique you'll want to consider during the recording stage of your project. For guitars, try using different guitars, amps, and EQ settings for your first and "doubled" track to give the result a bit more color than you might from a doubled track of the same setup.

Common guitar effects

Most guitarists are already familiar with these oldie-but-goodie basics, but we'll cover them here for good measure should you decide to apply them as you record (with the effects set on your amp, pedals, Pod; or digitally, after recording and in the mix stage):

- *Delay*—A hallmark of the Edge (U2), delay basically sounds like a gradually fading "echo." You can apply delay to any sound, but in rock records, it's often most prominently heard on and commonly applied to vocals, guitars, and drums to add atmosphere and/or sonic "depth" and power.

Continued

- *Chorus*—Gives guitar and vocals a potentially warm, melodious, stereophonic sound. Used extensively on Kurt's guitars for *Nevermind* and early Cure guitar tracks for songs like "Play for Today."
- *Distortion*—Can be generated by a physical or software amp or stompbox. "Distorts" the audio signal to give it a tough, jagged sound; used universally on rock recordings with electric guitar.
- *Flange or phaser*—Lends a trippy, psychedelic "wind tunnel"-like effect to sound, typically applied to vocals, bass, guitars, and synths.
- *Wah pedal*—A guitar effect that's controlled by a foot pedal that makes the guitar sound like it's "crying"—hence the "wah." Slash and Weezer's Rivers Cuomo often put this to good use, like on the solos for "Sweet Child O' Mine" and the intro solo of "Perfect Situation," respectively.
- *Tremolo*—Found on a lot of amps; makes guitars sound "shimmery" like ripples in calm water. Commonly used in "surf rock" by guitarists like Dick Dale. Also used for a "trippy," atmospheric sound in slower songs like those by the Cowboy Junkies and Yo La Tengo.

Acoustic guitars

These are generally recorded with condenser mics to pick up detail and nuances. Experiment with placement for best results, but you'll generally find the "sweet spot" with the mic aimed somewhere between the hole and the bridge of the guitar, about 6–12 inches away. You can also experiment with placing a second mic higher up or lower on the fretboard for more high or low end, respectively.

Keys

MIDI keyboards are generally recorded directly to your computer through your audio input device. Keyboard amplifiers can be recorded using the same techniques as described above for guitar amps. Keyboards that have audio outputs can also be recorded directly. For acoustic pianos, you'll get a reliable recording placing two condenser mics about 3–6 feet outside the raised lid, the mics evenly distributed between the lowest and highest keys along a baseline parallel to the piano. One mic capturing the overall sound will also suffice, but make sure it's placed well enough away from the piano to do so effectively. Acoustic pianos can also sound very good when miked from the back of the soundboard (uprights) and underneath (grand pianos).

Vocals

Here, you can use your trusty Shure SM57 or SM58, or a more expensive condenser mic—just always be sure to use the pop-guard covered in the Gear chapter to avoid plosives.

CHAPTER 7
Mixing

Before we get into more specific guidelines surrounding how to set up a basic rock mix, let's cover a few topics that are important to understanding the mixing process.

FUNDAMENTALS

Busses

Simply said, a bus is an audio pathway through which you can route multiple channels or input signals. Busses are most commonly used to group related instruments together to a single fader. For example, sending all your drum tracks to a bus (called a submix in this example) gives you the ability to adjust the entire kit's volume lower in the mix with one *fader* (see Appendix B, Glossary of Audio Terminology).

Effects sends are another kind of bus that allows you to feed, say, a reverb and apply it in varying degrees to a bunch of different channels. This can save you a ton of time in terms of not having to apply a favorite effect or effect combination to multiple channels—you can just route individual channels through the bus and adjust the degree to which those bus effects are applied. Using busses can also help your computer process more efficiently since instead of, say, applying the same four effects on four channels, you can route four channels through one bus.

Assigning individual tracks to route through a bus is pretty simple in most programs; refer to your manual for instruction.

Soundstage and panning

Soundstage describes the listener's "playing field" in terms of *localization*, the perceived location of a certain instrument or sound in the mix. A good stereo mix creates a seamless sonic illusion where the listener hears that the congo player is a few feet back on the (sound)"stage," the singer is "out front," keys are off to the side, etc. The reason a lot of beginners' demos sound flat and lifeless is because they're not creating this sense of dimension and space in their mix. Keeping your recording's soundstage balanced will make your mixes sound more dynamic, lively, and professional.

Every channel will have a *pan* knob that rotates about 270°. Keeping it set at 12 o'clock keeps the audio signal set squarely in the center, that is you'll hear it equally through both speakers. Set it a bit to the left or right, like 10 or 2 o'clock, and you'll hear the signal emphasized in the left or right speaker, respectively. Set it all the way to the left or right and you can eliminate it from either side completely.

One way to create a good stereo image is to use a bit of common sense and imagination by *panning* the individual tracks in your mix to match a physical soundstage you've envisioned for the song and its players beforehand (from the perspective of someone in the audience facing the stage). Generally, in rock recordings, bass guitar, lead vocals, snare, and kick drum are typically placed squarely in the middle, keeping their sounds "primary" and in both speakers in the stereo field. The lead guitar might be panned a bit to one side, while a rhythm guitar might be panned a bit to the other.

For rack and floor toms, you might consider mixing the highest-to-lowest-sounding toms from left to right, breaking the audience-perspective mix—that is, the leftmost rack tom from the drummer's perspective is panned leftmost and the floor tom is panned rightmost. This helps suggest a "sequential," "start-to-completion" sound like a tom fill "traveling" from left to right. You may want to create this start-to-completion kind of effect in your mixes with other kinds of sounds, too, like synth tones or other types of percussion where you want to suggest a natural distance and sequence between individual or groups of notes, sounds, or tones.

The more you experiment with and successfully manipulate panning and its effect on the overall soundstage, the more professional and lively your recordings will sound.

EQ

Humans can hear frequencies between 20 Hz and 20 kHz (Hertz and kilo-Hertz). Depending on your settings, adjusting the *EQ* (shorthand for EQualizer) of a song or individual track can emphasize, de-emphasize, or drastically alter the overall character of a sound. Applying EQ can make a snare drum sound much bigger or "fatter"; it can make a vocal track sound like it's coming through a telephone and it can help make a distorted guitar track stand out above the bass, or vice versa. Again, what's important is to apply it judiciously while making sure it's still somewhat pleasing to the ear.

With software EQs, you'll see that you can adjust multiple frequency ranges, or *bands*, on a graph. The frequency spectrum is represented from left (lower frequencies, starting at 20 Hz) to right (higher frequencies, ending at 20 kHz) on the horizontal axis; positive to negative dB (or "volume") is represented on the vertical axis (0 being the baseline).

There's some EQ-related terminology you should be familiar with, depicted in Figure 7.1 below.

EQ terminology visualized

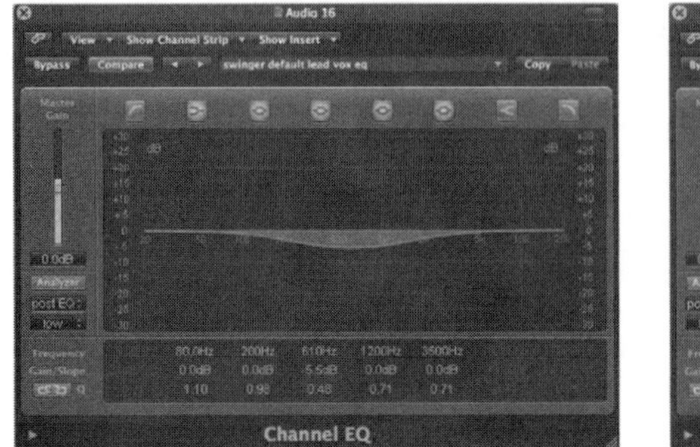

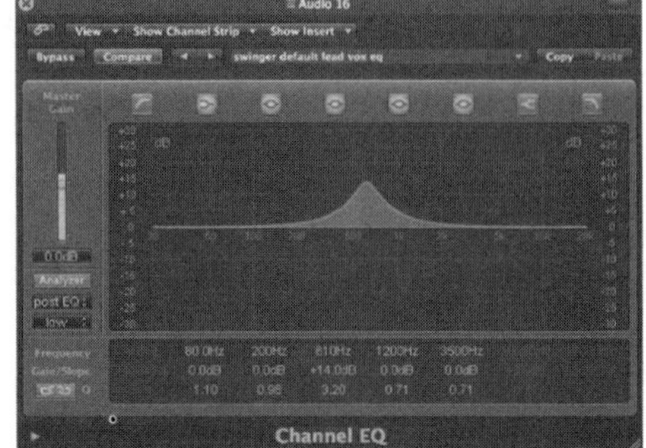

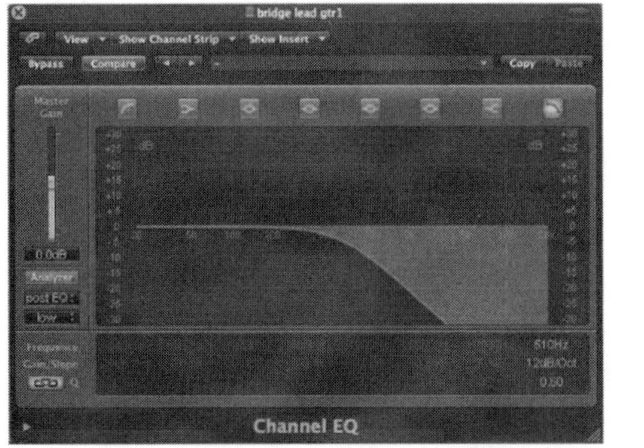

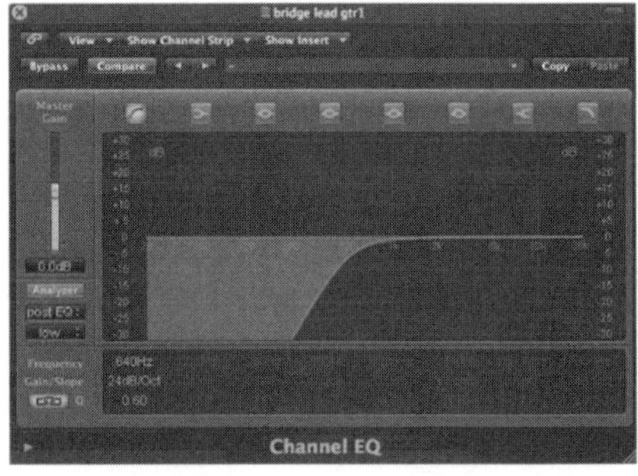

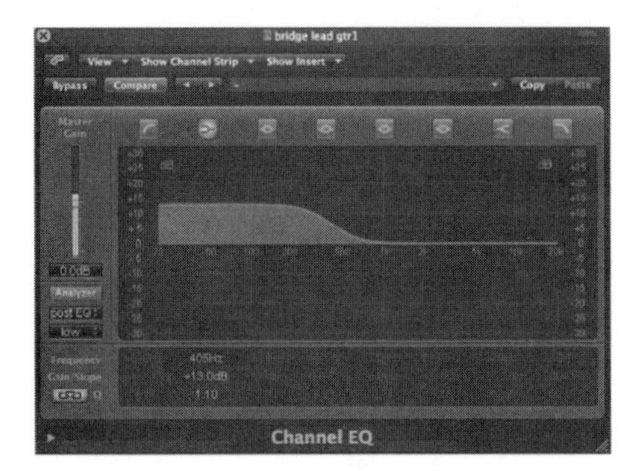

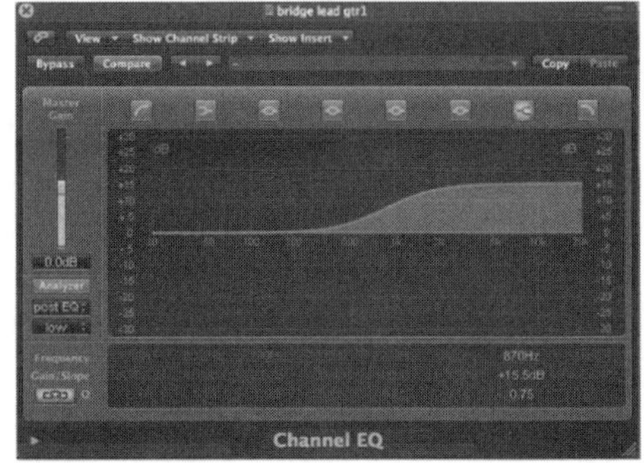

FIGURE 7.1
Single bands are represented here to help exemplify the terminology only; most digital EQ plug-ins allow you to set several bands across the frequency spectrum at once. Explore your EQ plug-in's presets (if it has any) for good starting points.

- Making a *cut* in a certain band simply refers to bringing it down (expressed in dB). To "roll a bit off" any band simply means to adjust that area down (e.g., "For rhythm guitars, you'll typically want to roll off a bit of low end so it doesn't compete with the bass.").
- Making a *boost* means to raise a certain frequency range.
- A *low-pass* (aka hi-cut) filter maintains lower frequencies but starts attenuating (or reducing in gain) higher frequencies at a certain cutoff frequency.
- A *high-pass* (aka low-cut) filter maintains higher frequencies but starts attenuating lower frequencies at the cutoff frequency.
- A *low shelf* filter boosts lower frequencies up to the cutoff frequency.
- A *high shelf* filter boosts higher frequencies down to the cutoff frequency.

For peaks or dips, an EQ band's *Q setting* refers to how narrow or wide you make its frequency range (e.g., a narrow Q setting looks like a sharp spike or crevice whereas a wide Q setting looks like a mountain or valley). The greater the Q value, the narrower the EQ band. This parameter can also be called "octaves" where the number expresses the number of octaves being affected. Q also applies to both shelving EQ and low/high-pass filters, where adjusting it affects the steepness of the filter slope (the higher the Q, the steeper the slope).

A common technique for isolating (i.e., finding) a sound's frequency range—and the area to adjust if necessary—is *sweeping*, where you simply drag an EQ point around the frequency spectrum to extremes (and to extreme high and low dB) as the track or song plays. For example, if a guitar sounds too trebly, you might sweep its EQ to find out where to cut or boost certain frequencies. It's important to note that it's often better to cut rather than boost when possible. This improves headroom in a mix, and in analog systems, improves phase coherency of the affected track.

Although it's common for more experienced recordists to apply EQ to individual tracks and across the master, newer recordists should try to keep it simple by making sure they're happy with the natural tone and character of the sound they're recording—that way one doesn't have to make a lot of adjustments afterward in the mix stage with EQ and other effects. Even the pros use EQ and most other effects as chisels, not sledgehammers, to help "sculpt" their sound.

Compression

A hardware or software *compressor* is a tool that's used to "compress" a dynamic audio signal to a more consistent relative volume. If you've ever wondered

FIGURE 7.2
Logic's compressor plug-in. Threshold, gain ratio, attack, release. Not too tough to get the hang of compression when you start experimenting.

why TV commercials are so loud, it's because the audio track's dynamics (the loud and soft passages) are "slammed" (industry parlance for applying a lot of compression) so that even the soft passages are as loud as they can be. Any compressor you purchase as part of your main software package or a plug-in should have documentation on how to use your particular compressor, but standard parameters and operation are similar (Figure 7.2).

Compressors become active based on an audio level threshold set by you; that is, they kick in when the audio signal hits a certain level. You also set a *ratio* (e.g., 3:1) of *gain* (signal amplification, basically) applied to the signal, which determines how much compression the audio signal gets when the compressor kicks in. Other standard compressor parameters include *attack* and *release* settings, which determine how slowly or quickly the compressor kicks in or backs off, respectively.

It may be clear by now that there are no hard and fast rules regarding compression, but there are a few general guidelines to follow. The first thing to keep in mind is that, like most other audio effects, compression should be used sparingly. A lot of rock producers and engineers are either currently supporting or resisting industry "volume wars," where the artists and labels often want their CDs as loud as possible—with producers and engineers over-applying compression to individual tracks and the master to get them there. Although using compression can certainly help a record sound louder, it sometimes comes at the expense of the overall dynamics or sound quality. More experienced mixologists quickly recognize over-compressed music as sounding loud but flat or "squashed," and certainly unnatural. Another odd symptom of over- or misuse of compression is an audio phenomenon known as "pumping," where the music seems to bob up and down volume-wise with the tempo or certain dynamic passages throughout the song. This is sometimes imperceptible to beginners, but as you become more experienced, you'll come to hear it easily and know when you need to back off on the compression.

Don't worry if you don't get the hang of compression at first—so many of today's digital plug-ins come with good presets for certain vocal, instrument, and song types and genres that you'll almost always have a good point to start and tweak from.

Compression is a great modern tool for helping to boost and control vocals, guitars, drums—just about anything, really. But use it sparingly.

SETTING UP A BASIC MIX

Here we'll cover what to consider with regard to individual sounds and their assigned channels: their typical volume in the mix, where they're placed on the soundstage (panning), and which effects may be applied (EQ, compression, delay, etc.)

The following guidelines are just that—guidelines help you to get started on setting up your mixes. As with just about everything in this book, everybody has their own techniques and preferences, and with enough practice and experience, you'll develop your own, too (Figure 7.3).

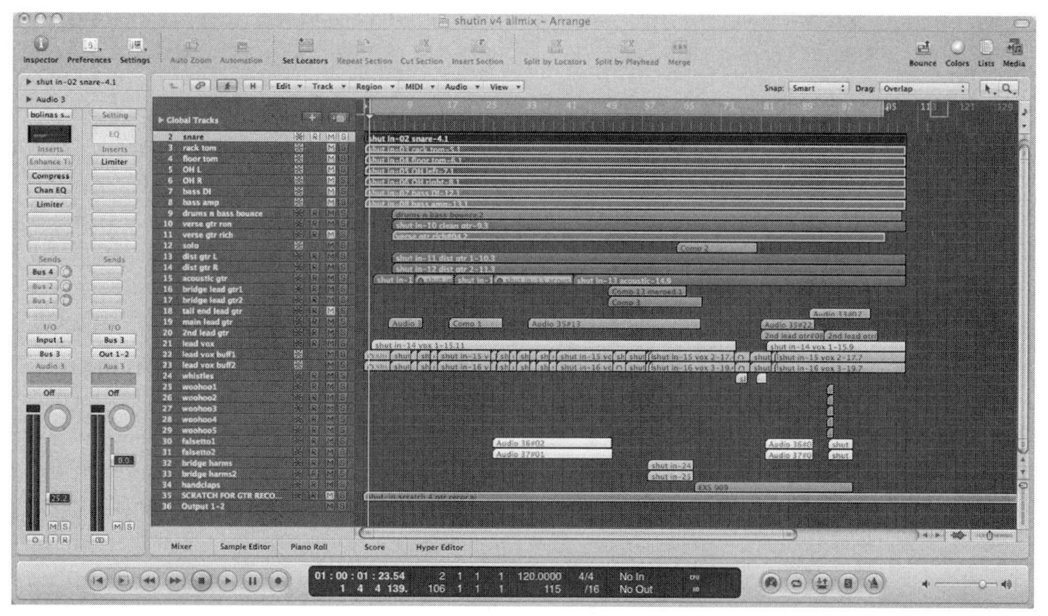

FIGURE 7.3
Screenshot of a mix from Logic Express. Everybody has their own preferences for setting up mixes, but establishing a template to use for new mixes will drastically speed up your workflow.

Drums

In *rock* mixes, it's fairly typical to set the kick and snare first since those are often the "loudest" individual channels around which the mix is built. (Don't forget to leave 3–6 dB of headroom above the channel you want highest in your mix, even if it's not the snare or kick.) I often find that whether or not the kick is set higher than the snare depends on the song's style and tempo: heavier drum mixes are good for slower rockers while lighter mixes work best for faster songs. *Tip*: A good rule of thumb when starting a mix is to set the kick's peak on the stereo output meter at around –18 dBfs. Adjust playback to a comfortable level, and mix the rest of the production around that.

PLACEMENT

If it's a four-track setup, with the snare, kick, and two overheads, it's common to pan each overhead to the left and right, respectively (I prefer about 10 and 2 o'clock myself). The kick and snare stay in the middle (12 o'clock, default setting). As already seen in "Recording", Chapter 6, with tom channels in the mix, it's common to evenly pan them individually from left to right so a tom roll dynamically "rolls" from left to right in the soundstage. For example, you might pan the left-most rack tom (left from the drummer's point of view) to the left stereo channel, a second rack tom a bit to the right, and hard-right pan (like 4 or 5 o'clock) a floor tom. This creates and exaggerates the sonic "illusion" of the physical left-to-right tom roll while keeping the overall drum mix balanced, lively, and fresh throughout the song.

EFFECTS

Apply reverb sparingly to drums, especially the kick and overheads. Most rockers like to add some compression and even some delay to drums (except the overhead channels, which should remain pretty dry), but again, don't commit the beginner's mistake of overdoing effects on everything—especially drums; it quickly muddies any mix. To add some depth and power to your toms and snare, try applying some gated reverb (reverb that "cuts off" at a set point, usually expressed in milliseconds; listen to Phil Collins' classic tom roll in "In the Air Tonight" for an example of gated reverb on toms).

Bass

PLACEMENT

It's fairly common to keep the mono bass channel "centered" at 12 noon, maybe a little off to the side. You'll want to try to arrange your mixes such that you can distinctly hear each instrument, and bass is no exception.

The challenge is usually arranging your mix in a way that other instruments aren't also interfering with the bass guitar's lower frequencies. The lower end of the frequency spectrum can muddy up pretty quickly (so can the mids, or mid-range, for that matter), so one way to prevent this from happening is to roll some low-end off your electric guitars and/or vocals, especially if you're doubling or tripling guitar tracks.

EFFECTS

Apply reverb to bass tracks sparingly (if at all), otherwise it loses its punch. Every bass guitarist is going to have a preferred sound, so if he hasn't already recorded his live bass rig sound replete with amp and effects coloration, you're free to experiment by adding a bit of distortion if that's right for the song. If you've recorded completely direct, or maybe only through a preamp with some EQ and compression going in, you can set the channel to a software bass amp modeler. I've found that these often sound just as good, if not better than most live rig recordings, so it's always a viable option.

Keys

PLACEMENT

Relative volume and placement are largely determined by preferences and the type of song. In guitar-driven rock, keys often play a supplementary role, where they might be used to add some interest and depth to the overall song. For example, a limited-note software *pad* (synth tone or chord with predefined character) might be mixed very subtly in the center or off to the side to help create a sense of depth and urgency to choruses, maybe saved for the last. If it's a keyboardist playing throughout the song, you can place the track in the center, or, more commonly, a bit off to the side. Obviously if it's a solo or a Norah Jones/Tori Amos kind of track, where the keys are integral to the artist's sound and their songs, it would be placed more "out front" and in the center. For pianos recorded with two mics—one for the low keys and the second for the highs—it's common to pan them out a bit left and right, respectively.

EFFECTS

Since synthesizers are often used because of their presets and effects, they often don't need effects other than perhaps some reverb, delay, compression, and/or EQ. If you recorded MIDI notes, you can of course change the sound, add effects, and even the character of the individual notes in the mix stage. For piano and organs, you'll want to leave the original character of those instruments largely intact, perhaps just add some reverb and compression, and place them in the soundstage in accordance with your preferences and in a way that's appropriate to the song.

Guitar

PLACEMENT

There are infinite options here, but let's cover a typical guitar mix for a basic rock record. As we covered earlier, you can add some instant beef to your rhythm guitar sound by doubling (or tripling, etc.) guitars, hard-panning one guitar track to the left and the other to the right. You can put lead or "background" guitar tracks in the center and off to the sides, respectively—just be careful that unless it's a solo, they're not competing with the lead vocal track(s).

EFFECTS

We covered guitar effects earlier, so if you've recorded a fairly dry or direct signal through a Pod with just amp modeler or distortion, now's the time to think about adding some reverb and/or delay. Adding a bit of delay to your lead guitar (especially solos) and vocal tracks can lend them a "bigger" sound—as if they're echoing in a large club or stadium—but again, use sparingly.

Vocals

PLACEMENT

Again, infinite options, but the lead vocal track is typically placed squarely in the center or slightly off-center, and not surprisingly, often set at a relative volume that's above everything else. Secondary or backing vocal tracks can be placed off to the side(s) to a degree that sits best in the overall mix. You'll find that doubling your lead vocal and keeping them both set in the center around the same relative volume does a nice job of "sweetening" the lead vocal track, which can sound good throughout or just to beef up the choruses. (You'll also want to experiment with setting the secondary track at a lower volume than the lead, and panning while your mix is playing to hit a "sweet spot".)

EFFECTS

Another great trick for vocals is applying different effects to your doubled tracks. For example, let's say you record your lead vocal fairly or completely dry going in—same with the doubled lead vocal track. In the mix stage, you keep that primary lead vocal track pretty dry, adding maybe just some light compression, EQ, reverb, and pitch correction (more just ahead on pitch correction in the "Plug-Ins" section of this chapter). For the secondary or "buffer" track, you keep it a bit lower than the primary vocal track but add a bit more reverb and perhaps delay, and no pitch correction to help the overall vocal stay sounding in tune but still adequately "humanized." You'll find that experimenting around this primary/secondary treatment on vocal mixes sounds good on your typical rock song, where fluctuating verse/chorus dynamics are common and it's prudent to add a bit of depth and power to lead vocals during the choruses.

When done correctly, these small, subtle tweaks in your vocals mix shouldn't bring attention to themselves but rather help create a slicker sound while sounding transparent to the average listener.

Now that we've run through how to set up a basic mix, let's take a look at…

COMMONLY USED AUDIO EFFECTS AND PLUG-INS AND THEIR APPLICATION

Plug-ins are software effects used with just about any of the better known audio production software packages like Logic, Reason, Pro Tools, Digital Performer, Cubase, etc. Most of these effects started as analog gear in the real world and are today emulated as software.

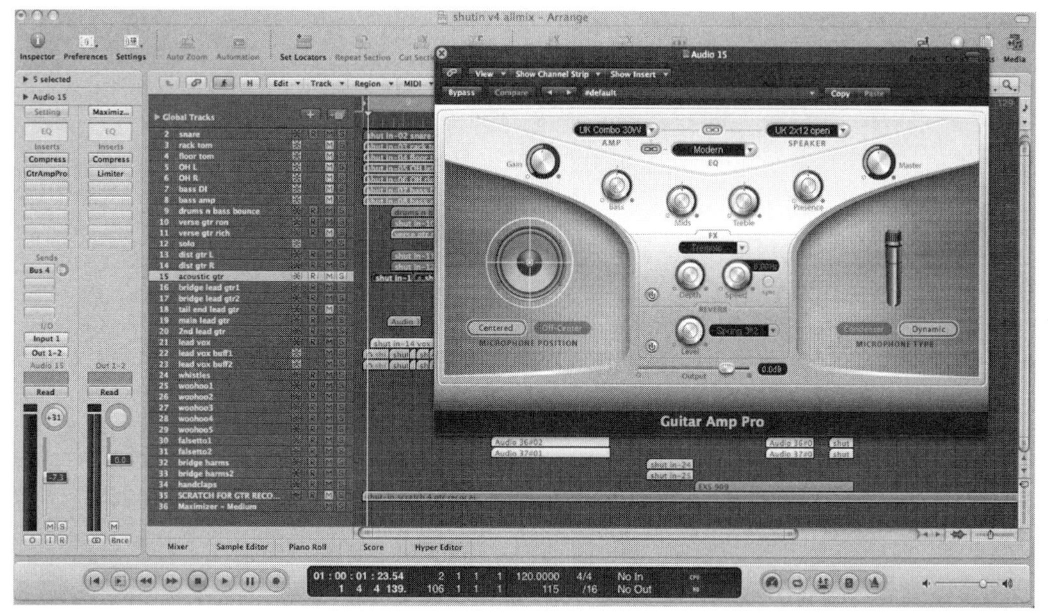

FIGURE 7.4
Logic Express's Guitar Amp Pro plug-in.

Plug-ins can range from the basic to the complex, from single-task to multilayered. What makes them so versatile is that even if you're working on a single software platform like Logic or Cubase, you can choose from any number of third-party vendor plug-ins to run your tracks through, from more common effects like reverb and delay to more sophisticated audio sweeteners like advanced pitch correction or even a mini mastering suite like Ozone, which includes multiple mastering-related effects in one plug-in.

There are different formats of plug-ins: Audio Units (for Apple) and VST (most common), RTAS (for use with ProTools), and Direct-X (less and less used today) being some of the better known. There are no right or wrong plug-ins to use, but they will have to be compatible with whatever main program you're working in. The graphic interface is also an important consideration: When you're mixing using more complicated plug-ins for hours on end, you want this kind of software to be easy on the eyes and easy to navigate, the two often going hand in hand. You'll also want to consider the price, and of course, how they sound, those two *sometimes* going hand in hand. The truth is that although some plug-ins most definitely sound better than others, don't put off recording until you procure that ultimate plug-in you've been coveting. Again, the best recordings are often the result of solid know-how, good material, and good performances, not a $1,000 mastering suite.

Some common plug-ins you'll likely be using in the mix stage include some of the guitar and vocal effects we covered in "Recording", Chapter 6, along with those we can categorize as being part of the mixologist's typical set of "clean-up" tools:

Limiter

A limiter is a type of compressor that puts a set "limit" or ceiling to the audio level. Like most plug-ins and audio effects, limiters can be applied to an individual track or overall song, helping to control those displeasing audio spikes and peaks while keeping everything under 0 dBfs (…and in the digital domain, you'll want to make sure every channel is below 0). The overall effect of well-applied limiting is a track that's more "even"-sounding and listenable.

Gate

Gates are used to help block out unwanted noise on a track or song like guitar buzz or a vocalist shuffling around in front of the mic. A gate "opens" to let the audio signal through when it hits a certain volume based on the parameters you set.

It's always best to optimize conditions before you start recording and mixing by making sure you're using the well shielded cables for guitars; the vocalist isn't shuffling lyric sheets or slapping his thighs in time with the beat; you lop off any noisy *heads, tails* (beginning, ends), and silent passages from individual tracks of your final mix. Granted, we can't always work under ideal conditions or with perfectly behaved gear, which is why we have tools like gates.

Pitch correction

Typically applied to lead vocal tracks so they stay in tune. This effect is often purposely *over*-applied in a lot of today's dance tracks in such a way that it makes the singer sound like a singing robot. That said, a little pitch correction can go a long way towards sweetening your tracks and making your songs sound a bit more "radio-friendly." As with most everything, less is more here: exercise good taste and don't quash the realness and intimacy of your vocal tracks by overusing it. Antares AutoTune is arguably the best-known pitch correction plug-in, but most main programs have their own version.

Time correction

Called "enhance timing" in Logic, time correction is typically applied to bass and drum tracks. As we pointed out above in the drum recording section, keep in mind that you can't use time correction if you haven't recorded to a click track, which sits on the same precise tempo map to which time correction makes its adjustments. You probably won't want or need to apply this effect to vocals or guitar parts that aren't more plucky or precise. With Beat Detective (ProTools), AudioSnap (Sonar), and others, the "on the grid" restriction no longer applies, and correction can be used on free-tempo tracks.

Like gates and pitch correction, time correction is one of those clean-up kind of plug-ins you should avoid relying on too much if you've properly recorded your tracks. To that end, make sure to rehearse, rerecord, and punch in your drummer and bass player (and any other band mates with tempo-sensitive parts) so they play as close to the click as possible.

Gain

Same as an amp setting gain, lends amplification or volume to a weak signal.

Now that we've covered the basics, and before we get into mastering, let's cover a few general guidelines to keep in mind as you firm up your mix.

GENERAL GUIDELINES

Trust your ears

Ironically, one of the good things about just starting out in audio production can be how little you know—you aren't yet tainted or confused by all the available methods, magazines, opinions, and options. While this is probably a good thing at first, you'll eventually develop your ear and mix skills by being able to identify and express certain audio phenomena and characteristics and know how to take corrective measures when things sound off. To this end, lots of experimentation, experience, and getting a grasp of audio production terminology will help. What's important to keep in mind is that no matter what your level of experience, you have to trust your ears. After all, it's your record.

Less is more

The first thing most beginners hear after sharing their first mix with those with more experience is that they've used too much reverb, chorus, delays, flange, etc. Remember, less is more. It's the hallmark of the amateur or insecure to smother their singing and playing with a ton of effects. Just focus on your performance and arrangements and keep the audio Photoshopping to a minimum. People want to hear *you*, warts and all.

Every action has a reaction

It's important to understand that changing even one thing can affect everything else in your mix. Boosting the level on the kick drum might muddy up the tight, clean bass guitar sound you had going. Lowering the lead guitar might make the lead vocals sound too loud. Adding some reverb to a lead vocal track will likely lower the perceived volume, and you'll have to adjust the rest of the mix accordingly. Panning your rhythm guitar tracks more towards the center might start to drown out the drums.

Mixes are a delicate balancing act, and unfortunately—even if the band sounds relatively the same across all ten tracks of a record—you typically can't apply a formula or template across the board. The trick is achieving the right balance across the frequency spectrum, the soundstage, and the perceived volume of individual tracks in the mix. When it comes to audio production and especially mixing, patience, perfectionism, and a willingness to experiment are virtues.

Practice makes perfect

One thing to notice after spending several months with your mixes is how your ear "develops." You'll notice when things are down a hair too low, the slightest overuse of reverb, the subtle differences a cut or boost in EQ can make and just about any other subtle nuance that's off. Further tweaking and listening will also help you better recognize, control, and use to your advantage all the topics here to help you produce a more professional sounding record.

If you've finished what might be your first mix, or just your best one ever, congratulations. Now that you've got that raw master track or whole record in hand, it's time to explore the final-phase production process of...

CHAPTER 8
Mastering

This chapter will cover what mastering is, what it entails, and considerations that surround whether or not you should D.I.Y. or leave to a pro.

WHAT IS MASTERING?

Surf the web, read the trades, and try to find some basic guidebooks and you'll find that even the definition of mastering is elusive. In a nutshell, mastering is the final polish you give your audio tracks to make for an optimally listenable, consistent-sounding record. To do it well is an art and science—a combination of expertise, experience, and judgment—that makes experienced mastering engineers well worth paying. That said, mastering a record in itself is by no means a "magic bullet"; while no mastering engineer can make a bad mix sound good, they can make a good mix sound great.

The role of the mastering engineer has changed a bit since the days of analog. In the days of vinyl, they used to work with all kinds of bulky, expensive tools, and hardware like metal plates and lathes. Then the advent of digital technology put the tools in the hands of the proverbial workers: Now anyone with a computer and a few bucks to spend on plug-ins can at least try to master their own record. Most major software platforms already include at least basic variations of the kinds of plug-ins you need to master, like volume maximizers, limiters, and multiband compressors. If not, you can check out third party plug-ins like those by Waves, iZotope Ozone, Bias Peak Pro, and T-Racks. With mastering software, you do generally get what you pay for in terms of sound quality.

Encouragingly, although not a lot of the pros would be willing to admit this, some of the better in-software mastering channel (often labeled by musical genre like "pop" or "jazz," and/or by short descriptors like "70s analog") and plug-in presets can often deliver good results. Try them. Even if the "rock" preset isn't quite right, or the "pop" preset sounds better on your rock track, you can use a mastering preset as the foundation upon which you can make your own tweaks.

Keep in mind that even though mastering presets can be quite good, the process of mastering itself typically entails more than a single-step, uniform approach. So before we get into considerations that surround whether you want to give it a shot yourself, we answer the question…

WHAT DOES MASTERING ENTAIL?

This list could get quite long, and there is no stock mastering checklist or sequence. Sometimes a track doesn't need to be mastered at all. Save trimming the heads and tails and applying a little gain (or volume "maximizer"-type of plug-in), EQ, multiband compressor, and/or limiting.

That said, here's what mastering typically entails:

Sequencing the songs

Establishing the proper song order is important, an art in itself. Treat your record—even a 4-song demo—as a mini-album, making sure that one song flows nicely into the next, and that it's got the proper peaks and valleys. Also keep in mind that the record listening experience is a bit different than the live listening experience, so you might want to use your set list only as a start.

Gain/limiter

We've already discussed these plug-ins, but working in tandem on the master signal chain, they help keep the overall volume high while making sure it stays below 0 dBfs, resulting in a professional, even-sounding listening experience. Although you can visually gauge your master volume in your software, there is no quantitative tool that can ensure your songs will sound perfectly consistent in volume across your record. The use of compression also plays into each track's actual and perceived volume—and they are indeed two different things. In other words, just because two songs both hover near 0 dBfs doesn't mean they'll both sound consistent in volume. In the end, achieving consistent perceived volume from one song to the next on a record often comes down to a matter of having finely honed critical listening skills.

Adjusting the frequency spectrum (EQ)

If your mix sounds good, you're probably not going to need to make a lot of extreme EQ adjustments. Experimentation is always a good thing, but whether or not you make a few cuts or boosts in your master track's lows, mids, or highs often comes down to personal preference and the overall sound you're after. The key is reaching an end-state where the song sounds even across the frequency spectrum—that is, lows, mids, and highs are dynamic but relatively even throughout the song. Speaking of dynamics as they relate to EQ…

Compression

Multiband compressors divide a signal into individual frequency bands (kind of like an EQ), and then compress them individually. They help increase a song's perceived volume, and can smooth out problem frequency areas. Remember to use compression with a light touch as proper dynamics, space, and air help any song sound more natural and prevent *listener fatigue* (an oftentimes subconscious suffering on a listener's part when a track has unrelenting volume, a poor mix, and/or frequency imbalance). Remember that it's usually safer to err on the side of keeping your mixes lower so people can opt to turn them up, as opposed to squashing the dynamics and creating a barrage of hyper-compressed sound they have to turn *down* (or more likely, off).

Adjusting the stereo image

Even when you mix in accordance with conventional guidelines, achieving an even left/right stereo balance doesn't always happen. More sophisticated stereo imaging tools can ultimately monitor and lend to the master track a more even stereo image so they don't sound too lopsided on either channel (Figure 8.1).

Now that we've covered the typical mastering to-do list, it's time to ask yourself...

SHOULD I PAY A PRO OR DO IT MYSELF?

You should know after a few tries with some good plug-ins whether or not you're comfortable mastering your own record. But before you start, you may want to ask yourself some questions:

Are your mixes worth paying someone else to master?

Don't spend money on mastering sub-par mixes or recordings. A good mastering engineer can't make a bad mix sound good, but they can make a good mix sound even better.

Who's the audience for your masters?

You probably don't need to pay to master a demo for prospective band mates or friends if you're just starting out, especially when you may need

FIGURE 8.1
Mike Wells of Mike Wells Mastering in San Francisco says that people often ask how much difference a professional engineer and facility can make. His answer: "On average if you translate typical improvement to a letter grade, it can make the difference of taking B mixes to A masters, or even C mixes to A masters. The difference can amaze you." (Photo by Chuck Revell).

to do just some of the basics discussed earlier like limiting/maximizing and EQ. You will probably want a pro to handle any recordings you intend to sell or share for "professional" purposes like getting airplay or using them as demos for labels.

Do you need the help?

After you've been mixing for several days, weeks, or longer, it almost always helps to have an objective, professional set of ears listen to your record to help you make informed sets of decisions: aesthetic and technical. Aesthetically, they can weigh in on the sequence of your album, which songs are your strongest, which ones can be cut, which mixes should be redone. Technically, they'll know what to listen for and how to correct or color your tracks in a way that optimizes the mix, and they have the professional listening environment, tools, and experience to do it with precision.

You can learn more about the art and craft of mastering—and how today's software applies to it—by taking classes from a local university or audio school, experimenting with your plug-ins, reading books, or watching a pro work on your own record, asking them questions along the way.

Although you can and should try mastering on your own with the tools you already have, it's widely acknowledged that professional-level mastering requires advanced technical knowledge and expertise, along with the gear, software, and a properly designed listening environment for the most accurate reproduction of your recording. Once you take the preceding questions into consideration and get more exposure to advanced mastering tools and techniques, you can decide if it's an area you want to focus on or pay a qualified specialist to perform on your behalf: even very good D.I.Y. musicians and producers leave this very important final phase to the pros.

PREPARING YOUR MP3s FOR DIGITAL DISTRIBUTION

We'll get into design fundamentals and sharing and selling your music online in Chapter 11, but before you post your MP3s, it's important that you take advantage of the kind of rich encoding that digital music formats like MP3s afford the artist. Take the time to enrich your MP3s by selecting them in your library and collectively or individually choosing (e.g., in iTunes) File > Get Info (or command-I on a Mac). Start with the "Info" tab, but you can also copy your lyrics and artwork into those tabs.

For the Info tab, you might want use the Comments field to list your URL and perhaps even influences and bands you sound like so that new listeners can get a quick grip on what you're about. For the Artwork tab, copy and paste your album or song art from a design program like Photoshop into the white space. Try to design for 7″ × 7″, at least 150 dpi since iTunes and most other media players have a nice "show or hide artwork" feature. In iTunes, you'll also notice that when you click the smaller thumbnail, a much bigger, bolder one pops up on your screen, so make your design the best it can be.

OK, now that you've got your band together and demo in hand, it's time to get out there.

PART 3
Promoting Your Band

By now you've formed your band, practiced, and recorded a demo. Now it's time to promote yourself, which can entail:

- Establishing your look with design elements like your record or demo cover and photo
- Carrying that same look and feel and relevant content to your website(s) and online press kit (or "OPK")
- Playing out
- Making a video

This last section doesn't get into putting together mini-tours and cracking multiple markets, mainly because the overall goal of this book is to give you the fundamentals you need to form, record, and promote your music. It's enough of a challenge to be good enough to rise to the top of the heap of local bands, let alone conquer the nearest few cities. It also takes commitment beyond being a weekend warrior group that practices once a week or has commitments and responsibilities beyond music. That said, once you finish Part 3 and at least establish yourself locally, you should know where you want to put your focus and if you want to go above and beyond playing locally.

But let's not get ahead of ourselves. What you need to do now is get your promotional materials together, start meeting other bands and playing out, and get some airplay on your local radio stations.

CHAPTER 9

Making Your Record Cover: Graphic Design Fundamentals and Best Practices

Now that you've read through Part 2, Recording, hopefully you've made a great demo. If so, you're probably pretty psyched to start getting it out to the world. But before you do that, it's important to understand some basic design fundamentals and apply them to your record or demo cover. As cynical as it may sound, everything you do with your band is part of your "brand"—that is, what people associate with you and your music. Your band's brand is indeed defined by the clothes you wear, the clubs you play, and the other bands you hang out with, among other things—including your music.

You only have so much control over your look, and while you'd hope to be judged primarily on the merit of your music, the fact is that ever since TV and videos came along, popular music has pretty much been a visual medium. As a new band, one of the fastest ways to be ridiculed—or worse, ignored—is to start promoting your music and identifying your brand with bad design. It's worth noting that there are books and websites dedicated to both good and bad record cover designs (search "bad record covers" or "worst album covers," or use "best" instead). The "bad" cover art sites are priceless in terms of providing some laughs at the artists' expense, and you'll notice that the work depicted in many of the "good" cover art books still stand up as good record covers today.

Before we get into what constitutes both bad and good design, let's start off with some context. For better or worse, just as the digital age has given just about anyone the tools to record themselves, another set of digital tools like Photoshop has also unleashed a new generation of D.I.Y. graphic designers. Regardless of how good your music is, it's important to register that people primarily rely on *visual* cues to make judgments, and they'll do so instantly based on your demo cover, flyers, photos, logo, website, clothes, and other elements that constitute your band's image. This is why it it's so important that your band's visual elements like logos, photos, and especially demo and record covers (analog or digital) are competently designed.

George Lucas once said that anyone can write, but very few people can do it well. Graphic design is no different. As artists, many musicians possess some degree of design aptitude or sensibility. If you're good at it (and this has been validated by someone other than your mother), we certainly encourage you to do your own cover art. But if you're not good at it, please ask an enthusiastic friend or pro to put something together for you that's good, or at least solicit their feedback.

Regardless of whether you D.I.Y. or have someone else design for you, it helps to know some graphic design terminology and fundamentals to help articulate and refine any ideas you might have. And since all the arts are related to one another—especially music, writing, and design—you'll notice that a lot of design (or painting, or photography) fundamentals apply to production and songwriting, and vice versa. So don't practice pentatonic scales all day or skip the following…

GRAPHIC DESIGN BASICS

Keep it simple

The key tenet in effective design—especially for relative beginners—is simplicity. Less is more. Again, this comes down to following the same design principles pros practice in other mediums like print advertising: every page should have one main image or idea; practice the rule of thirds (search "rule of thirds" online); avoid clutter and use negative space; use design effects sparingly.

Make sure it evokes your band and music

It's amazing how many beginning artists put out flyers and record covers that are completely incongruous with their image and sound. If you aren't an inherently visual person, at least take a look at CD covers of major-label bands you sound like to give you some frame of reference to start with.

Understand fonts

Fonts are style of letters that text elements and/or copy (words) are set in. If you don't have a good grasp of what kind of font or fonts best convey the overall mood or feel that your design is intended to evoke, talk to a designer friend—he'll know a lot more about fonts than you do. A basic rule of thumb is to try not to use more than one or two different fonts on a record cover, and definitely try to stay away from quick-apply effects like bevels, drop-shadows, and rotated text. Finally, when choosing what fonts to use, it's usually best to err on the side of visual simplicity. *Sans serif* fonts like Arial, Franklin Goth, Universe, Swiss, and Helvetica lack the serifs (i.e., little squiggles and flourishes) found in visually busier fonts. It's amazing how many bands adorn their record covers and flyers with fonts that are all over the map. Remember, keep it simple.

THE OPTION OF NON-IMAGE

More on band photos appears in Chapter 10 but this is a good time to qualify something about your band's design elements, which include your band photos: Sometimes the best image is a non-image evoked by your music, and not band photos.

Now that's not to say that all the design fundamentals we just covered aren't important; it just means that while being an attractive bunch never seems to hurt a band's chances, there are other ways to visually define yourself that may actually better support the kind of music you do and the image you want to convey. Great music has transcended a band's looks since the beginning of time. Bands like Tool, Rush, Radiohead, and Pink Floyd did not make it on their pop-idol looks; they made it on the vitality and originality of their music. All of these bands use impressionistic photos and videos sans band members to evoke the vibe or mood of their music. Some more examples of bands that have used this "mystery image" versus the "pin-up" image approach very successfully include Sun Kil Moon, R.E.M. (masters of this approach), Jimmy Eat World, Wilco, and Portishead. So even if you're a more "video-friendly" or image-conscious act like Duran Duran, The Donnas, Korn, or Green Day, you might want to consider lending an air of mystery to your band and music with more evocative and/or associative imagery (see Figure 9.1).

Crafting your image is really an exercise in creativity, good taste, and self-awareness. Do you know what you, your band, and your music are

FIGURE 9.1
Sun Kil Moon's Mark Kozelek's publicity photo is a perfect extension of the mystery, intimacy, and artistry of his music (Photo: Nyree Watts).

about? Are you willing to be yourself but ask hard questions about what aspects of your image may or may not be working and adjust accordingly? The line between being genuine and pretentious in the name of "image" is a fine one indeed, so always err on the side of staying true to yourself. And although there's nothing wrong with Ryan from accounting forming a band with a few friends, you should try to avoid looking like him in your band photos. People don't wanna see their neighbors rock—they wanna see a *show*.

CHAPTER 10
Your Press Kit

In today's day and age, doing your own publicity can feasibly all be done on your computer and online. You don't need to waste time, money, and paper on photos, press releases, etc.—they used to get tossed in the round file (i.e., the trash) in the old days, and it's no different today. Although a MySpace page or Facebook artist profile is OK for social networking (more on this to come) or fans, it's a bit more professional to have a separate web space to speak directly to press and bookers. That's where an OPK (Online Press Kit) comes in. **Sonicbids.com** is a great service for this purpose, but of course bands with more web savvy and resources (even a volunteer web designer/fan) can always just create a separate "for press only" website and get more design flexibility.

There are tons of Internet articles and books out there on creating the perfect press kit, but all of them share some common components. The important thing to remember is to keep it clean and simple. Press outlets, bookers, radio stations, and labels have tons of this stuff to wade through every day, so make yours stand out by being crisp, clear, and concise. Try finding press sites for your favorite bands if you can get access (some have a "press" tab in their main site) or browse sonicbids.com for examples. Here are the typical elements you'll want to think about incorporating into your OPK:

GENRE

This book is geared towards those who would likely fall under the indie rock, rock, or alternative genres. With digital music becoming so Internet-centric and global in nature, you'll find that a lot of sites are allowing for alpha and beta genres, usually 2 out of 3. Avoid the temptation to be cheeky and list your indie rock band's genres as something clever like "Rock/Soul/Christian Rap"; remember that lots of people are scanning the web at any hour in search of music in the genres that they know they like. If you want them as a potential fan, Christian Rap becomes an in-joke the potential fan or helpful industry player never appreciates.

BAND PHOTO

This can certainly be a deal-maker... or breaker. Unless you're better known and/or good at this from a design and photography standpoint, try to avoid being arty or clever with blurry, over-retouched, or oddly staged shots—just make it clear, colorful, and unique. If you have a stylish friend, ask him to attend any photo shoot along with the photographer, if they're not the same person. Bring a second and even third set of clothes and change backgrounds to get three sets of photos from one short photo session. Use natural light outside so you don't need to fuss with lighting for too long. Also, regarding your photographer friend: if he's good enough to lend you his time and talents, be sure to let him know you appreciate it by at least buying him lunch or putting him on the guest list for your next show.

CONTACT INFO

Don't make people hunt for it—put it at the top of your letters and web pages.

SYNOPSIS/SOUNDS LIKE/INFLUENCES

If you're brand new, it's always nice to have a quick "mission statement" at the top of your materials, a la *Johnstown: A 4-piece indie rock band based in San Francisco; influences include Weezer, Wilco and Nirvana.* MySpace has done a lot for musicians and interested parties by making "sounds like" and "influences" part of their presentation templates. People tend to need to categorize before getting to know you, so when you're just starting out, it's OK to borrow some interest by listing these other bands. Your main task is to make sure your *music* sounds original.

Also keep in mind that what you *sound like* and your *influences* may not completely overlap. You may *think* you sound like a combination of The Pixies and Rage Against the Machine, but your music may be influenced by any number of musicians—or even non-musicians like comedians or architects, for that matter. Definitely listen to and try to leverage in your PR what other people say you sound like—you'd be surprised at some of the self-awareness you glean from people trying to pigeonhole or label you, which they will invariably, instinctively do with any new band. Even if you don't agree with what other people say about your sound, they may be more objective and accurate in their comparisons.

SOUND SAMPLES

Online and on your CDs: Post your best three MP3s and make sure they can stand on their own; people usually skip around. That said, if "demo song" 1–3 is listed from top to bottom, put your best one at the top link. If you're passing out demos to friends, you'll quickly know which songs are your "best," and your demo (whether it's on a digital press kit or CD) is no place for filler. Extended intros to songs have no place in a demo.

Also, don't allow people to download songs unless you are absolutely certain you are never going to sell them. If they're available for free once, they're not as valuable as a commodity. Most digital music sites allow you to make the "free download," "short play-only sample" or "full play-only sample." (More about how to prepare your MP3s will be discussed in the coming chapters.)

YOUR BIO

You aren't famous yet, so just introduce your look and sound without a lot of dense, cheeky copy about the no-name lineup changes you've navigated in the last few months or years. No one cares about that kind of trivia at this stage in your fledgling career; just say "hello" loud, proud, and in about 200–400 words maximum. It's also nice to make this copy "scanable" using bullets (most recent first) to outline "career" highlights a la:

- 8/23—Opened for Weezer on second stage at last year's Live 105 FM Loudfest
- 7/14—Voted a "Top 10 Indie Record of the Year" in the *San Francisco Chronicle*
- 6/3—Won San Francisco KROK's "Battle of the Bands"

...and so on. A common pitfall of beginning self-authored bios is trying to sound too relevant when you're not yet well known. Keep in mind that your band is one of dozens someone might review on any given day, so write a bio that sparks interest from complete strangers. A good rule of thumb is to be brief, honest, and clear. Try to avoid sounding clever, cute, funny, or ironic. Unless you're very good at this, it can fail miserably in writing and it's best to establish your tone and just get down to business.

Here's an example of a bio of the kind of group that might be reading this book. Notice the tiny press angles and "hooks" that make it different from your average band (ex child-star as a member; opening up for a recently signed band, suggesting they're up-and-coming):

> Three-piece indie rock band Juicy Snaps was formed by four Beverly Hills high school students in 2004. Peter, Josh, and Davis all shared an affinity for '80s pop metal, alternative media, and punk rock. Lead singer Josh played wacky next-door neighbor Bernie in the '80s sitcom "Michael and Co." Now full-time students at UCLA, Juicy Snaps just started playing out in L.A. and recently opened up for local up-and-comers The Satan Babies, recently signed to BMG. Their first CD, *Sweet William*, will be sold on their website in December. www.juicysnaps.com.

HOW TO TRY/BUY

Sounds as self-evident as listing your URLs, but it's worth emphasizing: you must make it as easy as possible for people to listen to and/or buy your records. Don't make them navigate your 42-page website to do this. If you have distribution through iTunes Store, services like its Link Maker can create a custom link for you, so your listeners can click right through to your records. In short, allow people to try/buy with one click and always make how to get there a short, prominent and, ideally, *visual* call to action.

YOUR WEBSITE

To thoroughly cover the art and science of web design would be beyond the scope of this book—or any single book, for that matter. If you're just starting out, try to avoid spending an inordinate amount of time playing around with constructing and updating your website (unless, of course, that's your bag, baby)—focus on your music. You may want to throw up a MySpace page and be done with it; a MySpace band page offers virtually everything you need to promote your band online except sales functionality, and excellent, easy-to-use new services like Snocap are helping bridge that gap.

That said, a lot of bands that stick together or have some web skills often make their own home page in addition to a MySpace page, usually so they can offer more content and have more control over that and the look and feel. If you have to build your own site, you can ask an enthusiastic friend to help, find someone to build it for you on craigslist.org, or learn web design and software yourself. Some of the better-known programs for beginners out there that let you build sites without having to learn code include Freeway Express, Microsoft FrontPage, Apple's new iWeb, and Macromedia Dreamweaver. Pros usually work with the gold standard of web design software, Adobe Creative Suite, which includes Dreamweaver.

Remember, if you're a new band, keep it simple and shallow (fewer pages with light content) instead of deep (multiple pages, tons of content). Remember that the site should be designed with both industry types and potential fans in mind so they can find what they need quickly and easily. You'll want to have some or all of the following things on your webpage, if not as part of a navigation bar, then cleanly laid out on your homepage:

- **Home** – Most websites are designed so that if you click on a link or the logo in the top left of the page, it returns you to the homepage. A nice, intuitive thing to do for your site's visitors.
- **News** – Updates about your band can be posted as blog-like HTML-text entries linked to photos and sites in new windows—always a good format so people can scan by date and you can update quickly and easily.
- **Shows** – If you're really touring hard, you might want to make this a separate page with live photos; otherwise you can probably safely incorporate upcoming show dates into your "news" links.
- **Media** – Audio, video, pictures. Maybe little thumbnails of your catalog to give people a quick overview of how many records you've put out. Create your own digital booklets as complements to any records, and always, always put quick links or buttons next to any catalog so people can sample-listen/buy on impulse.
- **Bio** – See above, same guidelines: Keep it short, interesting, and relevant.
- **Links** – This isn't something you need to feature prominently or on your homepage, but it's always nice to support your mates' bands with links to them.

- **Try/Buy button** – If you're in business and selling your music, it actually lends you credibility to *not* be shy about sprinkling these rather liberally around your site in contextually relevant areas (like your "media" or "music" page).
- **Sign up for newsletter** – There's no better way to get fans out to shows than by using email (and being a great band); put that "subscribe" field for email addresses and a "submit" button on your homepage. Be sure to thank visitors for subscribing and give them a link to return to the homepage. You can also auto-email thanking them for subscribing, but these can be annoying if you've already acknowledged that you've added them with the auto-"thanks" response on your site. Also, try not to e-mail your list more than once or twice a month; people can always visit your site for any new news and updates. (More on email marketing services ahead). In today's D.I.Y. music business, your email list may very well be one of your most precious commodities.
- **Contact info** – If it's not in the nav bar, definitely put contact info on your homepage.

Finally, if you're an active band, be sure to **update your website frequently**—at least once a month with news, upcoming shows, and photos. It's a fact that websites that change things regularly get more clicks and rank higher in search engine results. Give people a reason to check back to your site often. Some of those might include:

- **Multimedia** – You might include homemade music videos (next section), digital audio and/or video recordings of your live show, or free downloads of "B-sides" or rarities that didn't make your last record. You could also include video or audio of podcasts of your band being interviewed. Posting photos of your last show are also a great way to keep people coming back to your site.
- **News** – This is where a blog-like area on your site can really come in handy—it's a fast and easy way to post what you're up to on a regular basis. It's also great PR for you to have what's essentially your own "wire service" with headlines dedicated to your band's latest accomplishments and activities.
- **New merchandise** – You'd think it'd be record sales, but merch is how most bands make their money these days. Create a mini-store with stickers, T-shirts, and CDs (ideally with bonus or unique material that might not be available on your digital record version). There are tons of ways to set up your own "web store" or e-commerce store online, with PayPal being one of the more popular e-commerce services (more on these ahead).

CHAPTER 11
Sharing, Socializing, and Selling Your Music Online

By now you know the elements you should have on your MySpace page, website, and/or off or online press kits. Hopefully you've got the band and a demo together, and you're going to play out (covered in Chapter 12). So this is a good place to cover what you probably want to do most after you record: share and sell your music. The fastest, cheapest way to do that—and with the widest reach—is online.

Before we get into it, it's important to qualify that while the web puts the means of global distribution and publicity in the hands of you, the artist, the web is not a magic bullet for band exposure and success. It still can't replace the kind of publicity and buzz and opportunities you'll get by writing, practicing your instrument, making great records, playing out, and networking with real people. Bands are still better off *pushing* their music to fans, other bands, industry types, and press instead of tweaking their online promotional sites and tools all day in the hopes that medium will *pull* new listeners. It helps to think of your website(s) and the web as a medium in general as a media repository, storefront, and calling card that helps keep fans and generates new ones, but even established bands need to be active (i.e., writing, recording, and performing) and, most importantly, very, very good to drive traffic to it. (No one makes the "It all starts with you" (not the web) point better than Mark Kozelek; see the following sidebar).

Mark Kozelek is the singer/songwriter and bandleader of Red House Painters and more recently, Sun Kil Moon (SKM). He also had a stereotype-bending role as a fictional bandleader in the film adaptation of Steve Martin's novella *Shopgirl*, which also features a few actual SKM songs.

What was the best thing you did to attract label attention? (Networking with other bands/side projects? Focus on making great records? Constant touring?)

Honestly, we (Red House Painters) didn't do any of that. When we signed to 4AD in early 1992, it caught us by surprise—I hadn't even heard of the label or played a show outside of San Francisco. I did my own thing—purposely avoided seeing shows, because I wanted to focus on my own thing. That was what it was about, developing my own voice, my own sound.

Mark Eitzel from American Music Club became a fan, and gave one of our demos to a journalist, who passed it onto Ivo at 4AD. That was the only label that paid us any attention. But the best way to attract attention is to lock yourself away, learn your instrument, write; then get out and play as much as you can.

What's the best way to maintain creative control and make money at the same time?

(Regarding) creative control: you do what you want, focus, and don't let others influence you. Making money? I've made enough to get by over the years, to pay my bills, create my own label, but I've never compromised in the name of making money. I've licensed songs to Wal-Mart, Target, various films, and tons of TV shows, but I've never given them anything beyond what I've already recorded. I do what I'm comfortable with. It's good to push yourself, to do as much as you can, but not to the point where you have people nagging at you and telling you what to do all of the time. You just have to do what you're comfortable with. If you can make a lot of money being yourself, that's great.

What's the biggest myth or lie about being famous and making music for a living?

Probably that it's an easy life. People are misinformed. A lot of people think "How can you complain? You get to travel, meet girls, sleep in." That's true maybe on your first tour, promoting your first album. But there's a discipline that's necessary with staying in the business for a long, long time—like 12, 15, and 20 years. If anyone thinks it's easy, then quit your desk job, make 15 albums, tour for 15 years, then give me a call and tell me about how easy it is.

What's your proudest moment as a musician and why?

Probably getting my first record deal. There is a lot of doubt that surrounds you early on. People don't know what to think about what you're doing. Your friends, employer, family, and even the guys in the band have their doubts. So that first record contract is validating.

What advice would you give beginning and mid-level indie rockers who are committed to making a living at music?

Quit f*cking around on MySpace and Facebook and all of that. Rent a room, play your guitar, write, rehearse a lot, play as much as you can.

Now that your expectations of what the web can do for you have been qualified, let's look at some of the better known sites and applications you'll want to take advantage of online. Although some sites are best at one or two main things, there's increasing overlap in terms of what certain sites can do based on partnerships and because more simply offer more "one-stop" services to the independent artist. The important things to consider are what services you want, any associated costs, the site's traffic and, yes, overall "cool factor."

Most bands just starting out are well served by giving their music away to friends, family and of course, fans. **Social networking** sites like **MySpace** and **Facebook** are great places to start because you can build online "social networks" or communities of online "friends" (a concept most teenagers on the planet are already familiar with) and share your music with them in digital music players on your profile. MySpace also lets you sell your music directly on your profile page through their Snocap widget, which lets people sample or buy your music, paying out a royalty on MP3s sold. Because MySpace is so established and dedicated to the musician in terms of its overall artist profile template, it's probably the most obvious place for new musicians to get online and start sharing their music (and news, photos, and videos). Because Facebook started out mainly as a social networking site creating artist profiles and running them feels a bit more "auxiliary" than a more musician-dedicated portal like MySpace. That said, you can use ReverbNation's "My band" application to create an Artist Profile tab to your profile (Search "my band" on Facebook to get started, or get started from **ReverbNation**, covered ahead).

Next up are sites where the focus is on connecting with fans by posting your **multimedia**: **YouTube** and **imeem.com** are good places to start. Imeem takes YouTube's "have your own 'web TV' channel" concept a bit further by being a space more for artists to post videos, photos, and music (…not that there's anything wrong with 'grandpa235's funny dog-chewing-peanut butter videos).

Another site worth mentioning for bands that are looking for a more advanced, dedicated, expanded suite of **promotional services** is **Reverbnation.com**, which describes itself rather accurately as a "marketing solution for musicians." If that sounds rather all-encompassing, it is: the site offers web hosting, email list management, online press kits, and provides a way to link to any on- or offline retailers or **online music stores** (more on these ahead) you may already be set up with from your profile page and the Store section of your TuneWidget (music retailers include the iTunes Store, Snocap, and AmazonMP3, among others).

Before we get into selling, there are a couple more avenues for promoting your music online to consider. As is the case with the other digital services and promotional tools listed earlier, there's increasing cross-pollination and integration between online music stores, sites, players, and **music recommendation engines**. The advantage to getting your music in certain digital retailers and recommendation engines is that it's possible for these engines to recommend your music to listeners based on their listening habits. Some of the better known music recommendation engines include the "Genius" feature in **iTunes, Pandora.com,** and **LastFM.com**.

Another interesting phenomenon in terms of using the web to promote your music is **remixing**. Remixing used to largely be more a facet of club and house music but more mainstream bands seem to be getting into the act. Basically, bands can upload elements of their song, or "stems," for fans to remix. With the web being such an interactive medium, art itself doesn't need to be limited to one-way "push" communications anymore—now it can be a two-way collaboration between creator and listener. We see the same thing happening with information itself on sites like wikipedia.org, where everyone can contribute to site content. At the moment it seems to be tough to say if remixing's going to take off, but for now it may be a good thing to do as a way to involve your fans in your creativity and creative "conversation" with them. Some good sites to look into for posting your music for remixing are SpliceMusic.com, MixMatchMusic.com, and IndabaMusic.com.[1]

If you're starting to gain fans and feel you have a big enough audience to start selling your music online, some of the more established and popular **online music stores** or **digital music services** include, of course, iTunes, Amazon MP3, Napster, and Rhapsody (visit good ol' wikipedia for a good comparison on online music stores here: http://en.wikipedia.org/wiki/Comparison_of_Online_music_stores). Because the online landscape

[1]*Web site challenge: Make roses out of "stems".* Joseph Tartakoff, Chronicle Staff Writer. Monday, September 29, 2008. http://www.sfgate.com/cgi-bin/article.cgi?f=/c/a/2008/09/29/BUKM131BEN.DTL

Share, Socialize, Sell: Sites & Services to Start with

site	overview	social networking / "add friends"	sell digital music	web hosting	email list management	online press kit (EPK)	post multimedia	manufacture CDs	sell CDs online
myspace.com	Sign up for an 'artist page.' Showcase and sell (with Snocap) your music, post photos, receive emails, design custom online ads, send mass bulletins, post your tour schedule, post video and band info like bio, news, blog and more. MySpace has essentially everything you need to maintain and grow a band presence online, and it's free.	X	X						
facebook.com	Mainly a social networking site but you can use ReverbNation's "My Band" application to create an Artist Profile tab to your profile. (Search 'my band' on Facebook to get started, or get started from ReverbNation).	X	X						
reverbnation.com	Good "one-stop shop" for indie bands looking for distribution and promotional tools.		X	X	X	X			
imeem.com	Upload photos, music, videos (not just video). More for artists and musicians than normal people.	X					X		
youtube.com	Upload videos.						X		
constantcontact.com	Send and track emails to your fans (You can see what links they opened, who opened and when, who subscribed/unsubscribed and more). Reasonable fee options.				X				
sonicbids.com	Dedicated to hosting artist "EPKs (TM)," or Electronic Press Kits. You can do just about anything with an EPK that you can do with MySpace, but it's a bit more robust and you can easily enter contests. All this for a reasonable monthly price.					X			
cdbaby.com	Initially established as a service for indie artists to sell their CDs online, it quickly also became known as a great place to go to get you on a lot of the normally-closed digital distributors like iTunes. One of the best independent music retailers because it gets you on so many other digital distrubution sites while enabling you to sell physical CDs at the same time.		X						X
discmakers.com	Established, well regarded CD manufacturer dedicated to the independent artist.							X	

FIGURE 11.1
Tip: One of the best ways to find out about how good bands are using the web to promote and sell their music online is to sign up for their email newsletters.

changes so often, it's not worth itemizing every site out there and what royalties you get for sales—those terms are available on the individual sites. Ultimately, you'll want to choose the service that best meets your needs and preferences and has the best reach on any given month or year.

Now that we've covered some of the sites and services that'll help promote your *digital* music, let's cover…

MANUFACTURING AND SELLING CDs

If the web makes it easy for bands to share and sell, and fans to sample, buy or steal (let's acknowledge that great white elephant in the room) online, why would you want to manufacture a CD? CDs remain handy promotional giveaways for friends, family, and industry types who for whatever reason may not spend a lot of time, if any, online. You may want to manufacture CDs that include bonus material—whether it's bonus audio tracks, multimedia, or both—for fans. CDs are a great takeaway/impulse buy for new and existing fans after you wow a crowd with your live show. Just realize that manufacturing CDs can be expensive, and although it's certainly a fun, great experience for any enthusiastic band to put out a CD, you'll want to consider whether it's a cost you want—or need—to recoup.

If one of the more reputable and established CD manufacturers is **Disc Makers** (and you'll find many more here: http://cdbaby.net/picks/1.html), then probably the best known online CD retailers is **CDBaby** (cdbaby.com). For a small setup fee, CDBaby gives your fans a place to buy your CD online while making your tracks available to buy on a wide variety of traditional retail CD stores and digital music retailers, including cdbaby.com, iTunes Store, Rhapsody, Napster, and Amazon MP3. If you're trying to decide whether you want to press a CD or not, there's probably no better person to speak to surrounding questions than **CDBaby founder Derek Sivers**, interviewed here…

Is the CD dead?

Not yet. We're in transition times. A lot of people have iPods. But most still don't. A lot of people get all their music online. But most still don't. Don't forget the true fact that more people are killed by pigs than sharks each year. We just hear about the sharks because they're more newsworthy. So they're not reporting in the news that "millions are still buying CDs"—but it's true. If you just read the news, you'd get the impression that nobody is buying CDs, just as you wouldn't know more people are killed by pigs than sharks.

With CD and record sales in general falling, do you think recordings will continue to be a viable way for artists to make money?

Absolutely. Independent artists are selling better than ever. Maybe Mariah Carey's sales are hurt by downloading and piracy, but indies are on a more level playing field than ever, with more access to effectively reach people than ever before.

Like any business, artists need to ensure multiple sources of income. Be a multilegged table, not a one-legged pogo stick. You can't do only sales, or only concerts, or only licensing. You have to have as many as possible to ensure a steady foundation.

How much music should new artists give away and why?

Give give give. Get it flowing. Make it easy for fans to copy and give to friends. Get thousands of free copies swirling around the 'net. But also have it available for sale. Some will want to just buy it. Some will hear the buzz and only buy it, because they want to support you. As time goes on, and your fame develops, have new material that is only available for sale, now that thousands are excited about what you're doing.

What is the best advice you can give new bands in terms of how to gain new fans outside their circle of friends, or beyond their hometown?

In short, read the last couple of books by Seth Godin. He says it in better way than I could summarize in a couple of sentences here. His books are an easy, inspiring read, and you're foolish not to spend the $15 and one offline evening giving it your full attention.

Even though the web gives a band more power to establish a presence and distribute, would it be fair to say that label support is still critical to mainstream sales and exposure? If not, why?

If by "mainstream" you mean "media star," then yes. To sell millions and have the millions in payola for every corporate FM radio station in the world, you'll need the shady insider dealings at the major labels. If that's the world you want to live in, you need to sign your life over to them and understand they will be your boss and control your career.

Luckily many are choosing to just make a good living, say $100,000 a year, by performing, recording, selling, and licensing. And that you can do independently.

You can't do it all yourself. You need a team. But you can hire the team so that you're the boss, not them. That's the crucial difference. That's the definition of independent.

CHAPTER 12
Playing Out

Your band is formed. Your demo rocks. You've put the band on MySpace and you may have even built a nice website replete with free wallpaper and video content. It's time to start playing out.

The first consideration here is your age. If you or band mates are *under 21*, you likely won't be able to (legally) play bars and clubs with a liquor license. Even so, you have some very viable options in terms of getting shows. You can:

- *Arrange D.I.Y. shows*—This is a viable option even for *21+ bands* (booking 21+ shows covered ahead). You can arrange to play in a space like a church basement, local recreation center, or university lecture hall—maybe even a "gimmick" show at the local art gallery or laundromat. Non-club D.I.Y. shows are a tradition in the all-ages indie rock and punk communities because it's a great way to start your own scene and community, where like-minded bands grow and share their fanbase. And without the club taking a larger cut at the door, you'll make more money. There's less pressure to pack the proverbial house, and you'll get a less formal and more flexible timetable and a more attentive audience than you might at the local pub. Green Day, Anti-Flag, Nirvana, and Soundgarden all booked their own all-ages shows in nonclub venues before they were signed. They may take a little more work, but the rewards of D.I.Y. shows can be worth it.

- *Book "all-ages" nights* at local bars and clubs (More on booking ahead). Lots of clubs and bars have all-ages nights where they don't serve alcohol and book younger bands. Sometimes they'll arrange the *lineup* or *bill* (i.e., the list of bands playing together on any given night, usually 3–4 a night) themselves or leave it up to you, offering the club and night for a cut at the door. In any case, just about any town or city has all-ages clubs.

By now you're probably very anxious to start reaching out to clubs and venues. Before you do that, though, and regardless of whether you're an under 21 or a 21+ band, you'll want to start your booking process by first doing some local reconnaissance and:

NETWORKING

The rest of this section is written primarily with 21+ bands booking club shows in mind, but it's relevant to bands of any age.

The music game is no different than the business game (hence, the "music business") in that the fastest way to learn and go places is to network. Now, we know that this kind of corporatespeak makes a lot of creative types cringe, but it's all in how you frame and apply this concept. If you do this well, things will happen more quickly than if you try to learn and do everything from scratch.

Let's say that, for whatever reason, you just moved to a new city, or a bigger city, or a new town with your parents—or you've moved in a concerted effort to "make it" in a major town like Los Angeles or New York. You've read this book, formed your band, recorded a killer demo, and you're ready to play out.

Do some basic reconnaissance first. Grab the free local newsweeklies (national directory here: www.aan.org/alternative/Aan/NewsweeklyDirectory), visit some clubs and local bands that you can see yourself sharing a bill with. If you really like the band, offer to help record some demos, do web or graphic or photography, pass out flyers, or pinch-hit as a "session" player—the stuff they could do themselves but maybe don't want to or can't do well. Become a part of their resource pool; become an asset to them. See them regularly—they'll appreciate having a new fan. Most importantly, start passing out your demo with a personal introduction if possible. Start making connections. This is how the world works, and "It's who you know" is a cliché because of its inherent truth. Just be sure not to get psycho about it or make a nuisance of yourself. After all, the point is to trade them something so you can leverage their fanbase and local industry network to benefit *your* band (known as a mutually beneficial or "win-win" arrangement in the business world).

BOOKING CLUB SHOWS

Hopefully after building your press machine and before you start booking, you've remembered the Golden Rule: your live act needs to be tight (great look, chops, chemistry, songs, etc.) and *draw* (i.e., as in draw a crowd) to be asked back.

Some clubs are more discriminating, perhaps downright snotty, about what bands they book (or not). Others just want the bar to make money that night. Some are a combination of both. You want to start developing relationships with a clubs and bookers who ultimately like your music and who might even want to help develop you. They might even give you a residency, meaning a regular show every 4–6 weeks, sometimes more often. And maybe—just maybe—they might throw you a plum slot opening for a national act on a good Thursday, Friday, or Saturday night. If your band sounds good *and* you can draw a decent crowd, you won't be short of gigs. You never know what doors a good set on your part might open, so make every show count.

After you've networked a bit, start keeping a short list of those clubs you like that host anywhere from 50 to 300 people. When you're starting out, location is everything: make sure your first show is at a place that's central and accessible to your crowd. You can start calling clubs to ask for the booker and preferred method of contact (e-mail for most), or just drop in with a CD.

If you've got your live act and online press kit, MySpace page, and/or homepage ready, you're ready to reach out to bookers. Your email query might look something like this:

Hi *(Try to address the booker by name as opposed to saying "Dear Club X")*,

We're <band name>, a 4-piece indie alt-rock band new to the San Francisco scene. Influences include Nirvana, Weezer, and Green Day.

If you've played out before or have a common acquaintance to mention, this is the spot to leverage your credibility and experience by mentioning that here, early on.

We're looking for a show in September or October. Even though we're just starting out, we'll work hard to promote the show and get a good crowd out. All of us have been to shows at <club> and think it would be a great fit, so we're more than happy to try out on a weekday night.

You'll find MP3s, bio, and photo at links below. Thanks for your consideration.

<your name>, <your band name>
www.yourEPK.com (electronic press kit with song samples, influences, bio, and photos)
www.yourhomepage.com
www.myspace.com/*yourband*

Band calendar (Shows dates already booked or unavailable):
Link to online calendar like Yahoo! Calendar or iCal in "month" view showing already-booked show nights and band's "unavailable" dates (such as when individual members are on vacation, or the band is out of town). This saves you and the booker a lot of back and forth.

Try to keep booking e-mail queries short and to the point. Bookers get tons of e-mail every day, so you'll basically want to cite any credentials you might have, along with anything else you might be able to do for the club, like:

- Get a good crowd out (cite an approximate figure if you think that'll help)
- Help find other bands to fill the bill (you're saving the booker the hassle)
- Extra effort by way of promotion, like any print ads you might be considering

Booking e-mails are basically business communications and should be treated as such, so if you're not the greatest salesperson or writer in the world, you may want to ask someone else in your band or an outsider to help out.

When you do book a show, ask the booker how the door is split (how much money you'll make, if any), when you should load in and when you'll go on, links and contact info for any other bands (so you can promote them in your e-mails and arrange to share gear or *backline*, i.e., gear on stage) along with any house rules (like no playing after 11 p.m.). If you've built a good band, demo, and promoted well, they will come.

Be prepared to send a lot of e-mails and make a lot of calls to get your first few shows. It's tough for new bands, but once you prove yourself a few times around town, it gets a lot easier.

THINGS TO KEEP IN MIND BEFORE AND AT YOUR FIRST SHOW

Bring a helper

A friend or two can help sell (and guard) your merchandise, take photos, and help you load in and break down. If this person is also good at audio or in a band themselves, you might also want them to listen to your sound-check and your set, (subtly) cuing you to any adjustments that may need to be made during the show. While speaking of sound checks, it's fine to request adjustments to meet your preferences, but try to let the club's soundperson do their job without too much interference. If you want your helper or your own soundperson to run sound, be sure to get the club's permission first.

Be cool to other bands

Don't be a prima donna and don't be a wallflower. If you're the band's leader, make a point of introducing yourself and your band to the other bands, and let them know to talk to you if they need anything (i.e., take responsibility as your band's representative, or assign one). If you're a live act, networking with other bands is probably the single best door-opener there is. If you've put together a good band—and you're cool to other bands—shows and opportunities will start coming to *you*. (*Tip*: It's both polite and beneficial to watch other bands' sets, even if you've already played).

Protect your time slot

Even when you're cool to other bands, it happens on occasion that they or the club may not be cool to you. You'll feel this if and when they try to change your time slot at the last minute. If this happens, and it's worth it to do so, remain polite but firm about wanting to keep your original slot. Changing time slots is bad for you and it's bad for your fans, especially if you've let them know you're going on at a certain time (and you should do this).

Always soundcheck first

Good soundpeople will want you to run through at least one song or partial before the show. Do your part to facilitate this by showing up on time, setting up quickly (and making sure you've discussed this with the other bands), and doing as they ask (when they ask for kick drum, everyone else stays quiet; stop playing your song when they say enough, etc.). And this cannot be stressed enough: be nice to your soundperson, and remember that you're not their "boss." Introduce yourself before the show, describe your players and gear and let him know what you'll need. A little courtesy and respect with soundpeople goes a long way. A little bit of attitude goes even further in the opposite direction.

FIGURE 12.1
Your soundperson (In this photo, Cal from San Francisco's Hotel Utah.) This is the first person you'll want to introduce yourself and be friendly to at your show. (Photo by R. Turgeon).

Prep the stage

Place your drink in a place it won't spill, and where you can reach it between songs (not on top of an amp). Tape down all of your cables so you don't trip on them and your gear doesn't black out during a song. If you're a guitarist, have an electric tuner on stage so you can tune up in between songs quickly. Distribute set lists for each band member. Make sure to place any backup gear (extra guitar, picks, snare, strings, etc.) onstage, so if something goes wrong, you don't have to amateurishly step off the stage to retrieve it.

Promote yourself during the show

It's rather surprising how few bands do this, but at a few choice moments in between songs, be sure to (briefly) mention your band, your website, any merch you might be selling, and your next show. It's also a good idea to thank the club before leaving the stage, give props to the opening band and a "coming soon" intro for the headliner (if it's not you). While emblazoning your band name or logo on your kick drum—or across a large homemade banner above the drummer—is somewhat hokey and gauche by today's standards, be sure to at least mention your band name and website. And don't forget to set a tablet out to collect names for your e-mail list.

Videotape and audio-record yourself

This isn't an exercise in vanity—it's a way to see and hear how you can improve. Professional sports teams review tapes of their games and talk about what went right and wrong, and you should make the effort to do the same. You may be very surprised at how different your act and music look and sound from your perception and memories of the performance. Video doesn't lie, which is what makes it such a valuable tool in terms of providing an honest assessment of your performance and how to improve it. Regarding audio-recording your performance, ask your booker and/or the soundperson if they can do this for you through the P.A.; some clubs have recorders on hand for this purpose. If you're gigging a lot, consider investing in an inexpensive portable recorder. Live audio can become great bonus or exclusive giveaway tracks for sale or as bootlegs; it's also good to review how you sounded. If you're spending too much time between songs explaining the hidden meanings of your lyrics or tuning up, you'll see and hear it all here—and hopefully won't do it next time.

Follow the rules

As we mentioned earlier, any club should tell you when to load in for the show. Although some clubs are more flexible about when you show up and when you can soundcheck, you should be there at the time they tell you to be there. If they have a webpage of rules and policies posted for

FIGURE 12.2
One thing bands should remember to do is make the effort to document their shows in photos and/or video. This photo is from a 56-page visual retrospective of photographer Ryan Schierling's 2007 and 2008 summer tours with Seattle's Spanish For 100. He self-published through MagCloud.com.

bands, read it. In music, sports, business, and life, there are rules, and one that pertains to all those areas is that it's smart to be punctual and professional.

Hot stage presence tips from Hot Lixx Hulahan

Craig Billmeier, aka Hot Lixx Hulahan, is the 2008 World Air Guitar Champion. He's also toured the world as a real guitarist in several real bands. Before that he was the 2006 and 2008 U.S. National Air Guitar Champion. Regarding your live shows and stage presence, memorize and live by these words of wisdom from the U.S. Ambassador of Airness…

I have performed on six continents in every conceivable venue, from soccer arenas to punk rock basements, bomb shelters in the Middle East

FIGURE 12.3
2008 World Air Guitar Champion, Craig Billmeier, aka Hot Lixx Hulahan (Photo by George Nitikin).

to $250/ticket jazz clubs in Hong Kong, cruise ships to elementary school daycares. And despite the vast diversity of these shows there remained a constant, one reliable and inevitable consistency: *audience apathy*.

If you're booking your own D.I.Y. tours then chances are good that until you start playing, audiences at large will not be familiar with (or remotely interested in) you. Accordingly, I have found that the single most powerful weapon against this indifference is putting on an energetic and impassioned live show—a spectacle. Give the audience something to remember; that is, why you are up on stage and they are not. Treat each performance as a physical and emotional catharsis. Even if your songs turn out to be a garbled, un-tuned mess—but you genuinely sell it—your audience will respond in kind. The Beatles learned that a few well-placed hip-shakes caused the girls to shriek so maniacally nobody ever noticed when they fumbled through a difficult passage (e.g., check out any performance of "Paperback Writer").

That which entertains is a vast and varied subject, but in the area of live performance I can definitively offer three What Not To Do's:

- *Don't shoegaze*—GWAR is the only band with shoes exciting enough to gaze at, and even they don't waste their time doing it.
- *Don't look bored*—disinterest begets disinterest. Charlie Watts gets away with it only because Mick Jagger is there to pick up the slack.

- *Don't let broken strings/undone hi-hat clutch/sagging mohawk/etc. ruin your momentum*. Have some jokes, a cover song, or sword-jugglers on the ready while the problems get fixed. The show must go on.

Let's be honest: barring your mom and the creepy guy who always offers to carry your gear, people probably aren't gonna care about your music at first. Heck, there are probably people in your band who don't even like your songs. *But keep your shows fun and exciting and they will come.*

CHAPTER 13
Making A Video

Making a video is one of the most effective ways to show the world what you look and sound like, and today's video content sites like YouTube, imeem, and MySpace let you "broadcast" your video to as many people who want to "tune in" online at any time of the day, from anywhere in the world. You can also post your videos on online press kit sites like sonicbids.com, so club bookers and industry people can check you out without you having to set foot in your tour van or on a stage. With the cost of digital video (DV) gear plummeting at the same rate as other consumer electronics—along with the growing accessibility and portability of video content via the 'net and mobile electronics—making a video for your band is as viably D.I.Y. as it is a record.

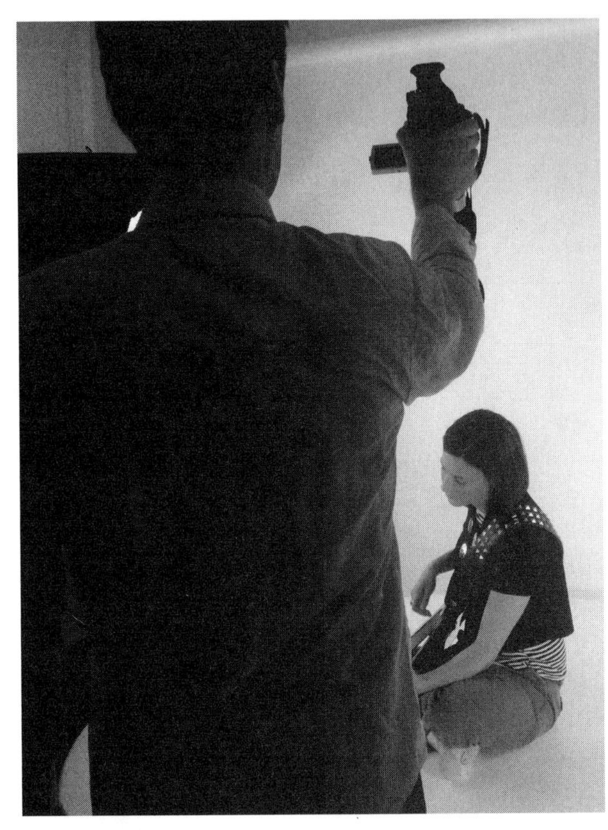

FIGURE 13.1
San Francisco's Tenderfew shoots a video for their song "See Me Sigh."

While making a decent video is definitely an acquired skill, it's not as tough to put together as you might think. I made the "Swinger" video (www.youtube.com/johnstownband) for the $80 it cost me to rent the lights ($40 for two shoots over 2 days in two locations). This section covers how I did it and the things you'll need and need to consider to make your own video.

WHAT YOU NEED

Camera or cameras

You can either rent a DV camera at a local photography shop or, preferably, use a friend's to save money. Having two cameras (and camerapersons) on the set to shoot the action simultaneously really helps get good footage from different angles that you can then sync up and cut between for a more sophisticated, polished "final cut."

An editor

This person could be you or a qualified friend. Although "anyone" can edit, not everyone can do it well. Like graphic design, photography, and production, editing is an art and craft and takes talent, experience, and skill. That said, it's in your best interests to work with an editor who really knows editing fundamentals and software like iMovie (which comes free with Apple computers), Adobe Premiere Pro, or Final Cut Pro (professional editing software packages).

Hardware

- Computer—For postproduction: editing, effects, and audio
- Lights—You'll very likely need to rent a light kit if you're shooting indoors. You can rent these for reasonable rates at most professional photography shops and suppliers.
- Portable sound system—You'll need to bring a CD player or iPod to play through portable speakers to lip sync to, if you'll be "singing" and "performing" in your video.

- Power strips and extension cords—You'll really need a bunch of these if you're renting lights. Err on the side of more, not less.

Mis en scene (and miscellaneous)

- Wardrobe—Plus make-up and ideally a make-up or hair person to work the band's style and, of course, cover up any unsightly blemishes or zits.
- Props—All the stuff that's not instruments or wardrobe that will appear in your video.
- Art direction and set design—Although the guerrilla video-maker will want to find a set that's "ready to go," it's amazing what a few touches and changes can do to bring life to any scene and shot, especially in the areas of lighting and color. If this "production designer" ends up being you, remember to keep an eye on everything that's in the frame and that in a visual medium, a picture is worth a thousand words. Always ask yourself what story and mood is your set is conveying.
- Instruments—You need 'em on hand if you'll be "performing" in the vid. Don't forget any cords, amps, pedals, drum sticks, etc. related to your live show that you'll want to be seen "playing" in your footage.
- Still camera—Great for documenting your shoot, and you can also use these stills for publicity purposes and your website.
- Shot list—If you're shooting a video concept that tells a story, this'll help ensure that you get all the coverage you need and don't waste time.

Again, gear and software is the easy stuff to learn—and there are manuals for that. What follows is a guerrilla film school crash course—the basics you need to know to get out there and make the video without spending a ton of time and money. And let's face it, if you've got some skills and creativity, you don't need to.

Video or movie shoots at any level are typically divided into three phases: **pre-production, production, post-production**. Let's look at what each entails…

PRE-PRODUCTION

The concept

The blueprint for any narrative film is the screenplay, which provides the action and dialogue. Unless there's a narrative "intro," music videos don't have dialogue—the song is the soundtrack. Whether it's a film or a music video, you should still come up with a *concept* or rough "story" to help engage viewers. Even if you just want to shoot your band performing, having a concept really helps set the video apart. The concept can hinge on a cool set or setting, or tell a "silent" (i.e., no dialogue, just through action) story, or do both simultaneously. A common music video convention is to split between a "story" concept of some sort and the band performing, usually in another location. But hey, it's your video—be as creative and imaginative as you like. Just try to maintain some sense of cohesion and consistency throughout the video.

Storyboarding/shot list

You don't always need storyboards or shot lists for simple shoots of the band just lip syncing or performing—just get enough "coverage" (i.e., footage) to make sure you can get a complete edit of the song together. But for videos with multiple locations and/or a "story" concept, putting together some *storyboards* (thumbnail illustrations that sequentially "tell" your story, like a comicbook without the panels and dialogue) and a basic *shot list* (a list of the shots categorized by date, location, actors, and other production elements and considerations) is a good idea. Even if you have the entire video mapped out in your head, even crudely drawn storyboards can help you communicate your vision to your band, cast, and crew.

Even if you're not telling a more complicated, linear story in your video, there are so many details to keep in mind on even a basic video shoot that you'll soon find that your shot list is probably the most valuable item on the "set" that day. It's insurance that you'll get your shoot done on time and won't forget to get the coverage you need, especially if you only have certain gear, people, and locations available to you the day of the shoot. You don't want to have to get everybody back together because you forgot to get close-ups of the singer (but shot everything else). For more complicated shoots, you'll definitely want to organize and use your shot list almost as a schedule, estimating the time you'll take for each location and camera (and lighting, if you have lights) "setup." Setting up your camera and lights shouldn't be *too* time-consuming for any low-budget indie video, but you'll want to factor that into your shot list/schedule, along with packing and setting back up if you'll be moving from location to location throughout the day (...not recommended—try to keep to one location a day to keep things simple).

Make a list, check it twice

Before you start shooting in the "production" phase, *make a list* of everything you'll need to physically bring, using the "Mis en scene" list earlier as a general inventory checklist. A forgotten DV tape or cable can waste an hour or even the entire day if you've trucked everybody out to some cool empty swimming pool in which you intended to shoot your band skating and/or "lip syncing." It gets worse if it's the only day you have to shoot at your friend's parents' pool…

OK, you've got your concept, storyboards, shot list, and inventory list together. Now it's time for…

PRODUCTION

Congratulations, you got everyone and everything to the set in one piece—it's time to start shooting. The important thing here is to make sure that you're using your shot list to stay on schedule, the lighting is good (if you're using artificial light, do some research on how to do this effectively) and that everybody's comfortable, especially those appearing on camera (they'll open up more and their personalities will come through on camera). Most importantly, keep in mind that filmmaking is a collaborative process with a lot of variables and moving parts: you will have to remain flexible and learn to make some compromises with your gear, band, and circumstances. One of the best parts about filmmaking is working with what you have—predicted or not, accidental or planned—and using your wiles (and the editing room) to make sure it all comes together in a way that works.

Important: When you're finished shooting, make sure your DV tapes are stored in a safe place, and if you can and you're the insurance type, make backups right away.

Finally, before we move into post-production, let's look at some time-tested…

Guerilla filmmaking tips (for the director)

- *Be generous*—Feed your crew and band. It's amazing how far a little food and drink will go in terms of boosting morale and reciprocation.
- *Keep your overhead low*—If you can do it for free, do it for free. Make it a personal challenge to make your video for as little as possible after any hardware purchases you or friends might have already brought to the table.
- *Get friends to volunteer*—As actors, extras. Offering a chance to appear on camera is enough to make a lot of people show and "work" unpaid for the day, but throw a little pizza and beer into the mix and watch how much more popular you become.

- *Fewer actors, fewer locations*—Low-budget, indie filmmaking rule #1: fewer actors and locations = less hassle, and fewer dollars spent. People want to see your band, anyway, so don't put too much priority on having multiple locations/tons of extras in your video if it's a hassle.
- *Use daylight*—Renting lights costs money and lighting a set is an art and craft unto itself. If you can shoot outdoors during the day, and especially at dawn (also known as "the magic hour" because of the naturally diffused, golden sunlight) or dusk, do it.

Insurance, permits, and liability, oh my

Stuff you won't like but might want to consider…

If you use a location that's owned by anyone other than someone you know, you may need to consider protecting yourself with insurance. In professional productions, using anyone else's location typically requires around a cool million in general liability insurance to protect the property owner from being sued if someone gets hurt. Blocking traffic and waving a prop gun around are two other potential but major no-nos without proper permits. All SAG (Screen Actors Guild) actors you might use would require that you purchase workman's comp. All of these considerations make it important to at least be aware of local rules and regulations with regard to insurance and permits (this would apply to D.I.Y. shows as well). A good place to start is your local film office; every state has one and you can just search "state film office" to find them (also try searching your city+"film office" or "film society" and you'll get plenty of places to start). If you live in a smaller town, a next-best resource is your local university (with a film department, of course) or production company (search "Production Companies" in the Yellow Pages, online or off).

OK, by now you've got the footage you need to piece it all together in the editing room. Which brings us to…

POST-PRODUCTION

Welcome to the editing room (i.e., your laptop or computer). The digital era has ushered in a new era of accessibility and ease with regard to shooting and editing your DV project—you just need a computer and some editing and audio software. If you're a Mac person, iMovie and GarageBand are designed to provide an integrated and easy-to-learn and use way to put together short video projects like music videos. PC owners can check out consumer-friendly video editing packages like Pinnacle

Studio, Cinematic, Ulead VideoStudio, and VideoWave. Final Cut Pro and Adobe Premiere are the two industry standard suites for more advanced and professional editors.

Like lighting (handled on a film set by a director of photography, or "DP"), editing can make or break a film or video, so know what you're doing or hire or recruit a good editor. Even if you have good ideas about how you want your video to be "cut," but you're not great with the technical side of things (e.g., the software like Final Cut Pro), it's always a good idea to team up with someone who is more technically proficient so they can cut as you "drive." If they're experienced and good, you'll come up with something better than you could have done on your own. And again, it's this collaborative "putting the found pieces together" aspect of music and filmmaking that can make them such fun and rewarding pursuits.

Although a "how to edit" is beyond the scope of this book, a good place to start is the "film editing" entry on Wikipedia, which, oddly enough, doesn't cover basic transitions from shot to shot. Such transitions are often easy to implement with today's editing software and can include the...

- *Cut*—This is just going from one shot to the next, no effects. Straight cuts or edits from shot to shot make up the majority of most of what you see in narrative and music film and videos.
- *Dissolve*—When one shot "fades out" into the next one "fading up" over some given time. Usually used to denote the passage of time or provide a "softer" transition between shots.
- *Fade out*—When a shot fades to black. This usually means the end credits are to shortly follow or it lets the viewer know the scene or video is over. In movies, the next shot (sometimes a fade-in or cut-to) usually takes place a few hours or days later.
- *Fade in*—...from black, usually denotes a "beginning" of a story or scene.
- *Jump cuts*—A postmodern, frenetic, disjointed style of editing that helps compress time by more quickly than usual cutting between related shots.
- *Wipe*—An editing "line" that can run left-right or top-down on the screen; used extensively in the *Star Wars* films, usually to transition to another location but to something that's happening "at the same time" in the narrative. Might be verbally expressed as "Meanwhile, at the Death Star..."

Editing transitions: The "tap" exercise

Now that you have some of the editing transitions/vocabulary down, it's time to learn a nice film school exercise that will help you get a sense of what editing's all about. (Note: It's nice to try this with your thumb on your DVD player remote's Pause button. If you have a DVD player or computer that can play DVDs in "slow motion," even better).

Put one of your favorite movies in (or even a music video) and perhaps skip to one of your favorite moments or scenes, preferably a key scene in the film. Once you're there, tap your index finger on some surface or your thumb every time there's a transition or cut. Because we're so used to editing conventions, it's very easy to forget about doing this after a minute or two, so really try to focus. After your scene and tapping is over, do it again, maybe 4–5 times during that key scene. Become very familiar with the rhythm and placement of the edits. Then sit back and think about these questions. They'll help inform how you make your own "cuts":

- How did the editing affect the mood or emotion of the scene?
- How did the edits move the story along? Did they seem to slow time down or speed it up, or both?
- Were some cuts "hard" (like straight cuts) or "soft," like dissolves? Was there a mixture?
- What was the "tempo" or pace of the cuts? Usually sad or dialogue-heavy scenes are cut with a slower pace, while action scenes are edited with very frequent cuts to keep the visuals and "movement" on screen exciting and dynamic. There's also more compressed "story" to tell through pictures in action scenes with a lot going on. Notice how the pace of the editing picks up when the story or action picks up or a new character enters the scene. Likewise, the pace of the edits tends to be much slower when the focus is on dialogue and character development.

CHAPTER 14
Sticking with it

By now, you've learned the basics behind running, recording, and promoting your band. Once you've played a few shows, met a few other bands, and recorded some songs, you're well on your way, and way ahead of those who never muster the courage, determination, and discipline required to get out there and rock. For anyone who loves it, music is a lifelong pursuit. I'll leave you with a few final thoughts on what it takes to stick with it, and maybe even make a living at it.

DO IT FOR THE RIGHT REASONS

If your main reason for getting into music is to be rich and famous, eventually you'll quit (unless, of course, you actually get rich and famous). Your chances of success are much better if you're doing it because you love it, and you can accept any personal or financial sacrifices you may be making along the way.

KNOW WHEN TO TAKE TIME OFF

If you're getting tired, miserable, or bitter, or all three, stop or shift your focus until you feel ready to get back into the game or you find a new path to pursue (musically related or otherwise).

NEVER STOP LEARNING

Congratulations, you're about to graduate from Indie Rock 101. But this is only the beginning, and with so much to create, learn, and explore, there is no end. Even if you only want to play drums or produce, the more you know about music, performance, production, promotion, and the business, the more it increases your chances of going pro. More importantly, it increases your chances of making great music.

Remember that every song you hear in your head is a gift, so get out there and perform, produce, and promote it—and do it yourself.

Appendix A Audio Production Magazines

- *Computer Music*—(www.computermusic.co.uk) Sister magazine to *FutureMusic*. Billed as "the complete guide to making music with a computer," features tutorials, hardware/software reviews, glossaries, FAQs, interviews, integrated DVD-ROM, and more.
- *Electronic Musician*—(www.emusician.com) Geared towards the serious hobbyist to semi-pro/pro. Skewed towards a more mature industry crowd, but it's a mainstay for many reasons, including excellent feature stories and interviews in particular.
- *EQ*—(www.eqmag.com) Nicely designed and a light but rewarding monthly read; rely on it for good production-related interviews, product reviews, and how-to articles. Better-than-average layout and design is easy on the eyes and easy to navigate.
- *FutureMusic*—(www.futuremusic.co.uk) For home or project-studio-based recordists at the beginner to serious hobbyist level, FM is one of the best-designed and written music production magazines available, consistently featuring timely, practical and highly readable recording, production, and music business info and trends.
- *Mix*—(www.mixonline.com) Aimed at a more pro-level industry audience. A bit scattered from both a design and editorial standpoint, but reliably full of product and record reviews, deep feature articles, "Classic Tracks" and "Coast to Coast" studio profiles.
- *MusicTech*—(www.musictechmag.co.uk) A sister publication to *FutureMusic*, aimed at producers, engineers, and more experienced recordists.
- *Sound On Sound*—(www.soundonsound.com) Lacks slickness in terms of layout, design, and cohesion of content, but always reliable and workmanlike in its quality, depth, and substance. Features techniques, rich feature stories, and tons of product tests every month.
- *Tape-Op*—(www.tapeop.com) Free by subscription, independently published 'creative music recording' magazine aimed at the indie, D.I.Y. recordist. Enamored with all things lo-fi and analog and infused with a refreshing commitment to D.I.Y. self-reliance, ingenuity, and passion.

Appendix B Glossary of Audio Terminology

Audiophile - Self-proclaimed audio "experts" who typically focus their energies on buying, experimenting with, listening to, and reviewing new audio formats and technologies (mainly home audio systems and speakers).

Bit depth - Amount of bits per sample, usually 16 or 24. Sample rate times bit depth gives you the bit rate of uncompressed digital audio. Greater bit depth translates to finer recorded volume resolution.

Bit rate - Usually describes the amount of data per second used by MP3, AAC encoding. Higher bit rates result in higher quality encodes. Can be used also to describe any digital format referenced against time.

Bottom end - Describes general presence of low frequencies in a song, system, or sound.

DAW - Acronym for "Digital Audio Workstation."

Delay - Audio effect that delays audio by a value—usually expressed in milliseconds—that can be mixed with dry signals to produce echo, slapback, flanging, or chorus sounds. Feeding back the output of a delay in the input of a delay can create thicker tone, rhythmic pulsing, or pseudo reverb effects.

Dynamics - Describes the overall "loudness" or "softness" of a song. A song with good dynamics strikes a balance between loud and soft, and between fuller and simpler arrangements at different points in the song. Dynamic range is a term used to quantify a device's capability of handling soft to loud signal levels.

Fader - One of those up-and-down "sliders" on hardware and software mixing boards that adjusts levels.

FireWire - An external data interface like USB. Can be used for audio interfaces, hard drives, DV cameras, and other external computer peripherals. Different "types" include 1394a, 1394b, iLink. Can be 4 or 6 pins at 400 Mb/s or 9 pins at 800 Mb/s.

Latency - The delay between input and output of audio through a computer. Also applies to the time lag between hitting a key on a MIDI controller, and the output of audio from a soft synth. Largely determined by a soundcard's buffer setting.

Near-field monitors - Speakers designed specifically for mixing music in close quarters, that is, they sit close to your ears.

Overdubbing - Any recording that happens to augment already recorded material.

Pad - Basically a synthesizer-generated chord or tonal "wash"; often used to add musical texture or atmosphere to a song.

Pan - Adjustment controlling where a sound appears from left to right in the stereo soundstage.

Plug-in - Software effects. Plug-ins range from simple EQs and compressors to more advanced effects like stereo wideners, adaptive limiters, and multi-band compressors. Typically, these take the form of VST, RTAS, AU, or DirectX effects. Your DAW program's compatibility requirements will dictate which is appropriate for you.

Pop guard - Typically a nylon or metal mesh "screen" placed between a singer and a mic; minimizes the harsh "plosives" (P's and B's, usually) by reducing the force of a vocalist's air blast before it hits the mic.

Preamp - Or "preamplifier"; any hardware that amplifies a weak input signal. Typically used for microphones, guitars, and basses.

Preset - A setting on hardware or software that's "preset" by the manufacturer. Very helpful in terms of establishing a baseline from which you can adjust. For example, a software EQ might have a preset that makes vocals sound like they're filtered through a telephone line and a hardware synthesizer may have presets that emulate any number of real-world instruments, like trumpet or piano.

Quantize - A MIDI function that automatically sets notes perfectly in time to parameters set by you.

Sample Rate - The number of samples or "snapshots" of an analog signal taken per second during the digitizing process. For example, 44,100 times per second is the sample rate of CD-quality audio. Higher sample rates allow higher frequencies to be recorded.

Soundstage - Describes where certain sounds, instruments, and tracks appear in the stereo field.

Subwoofer - A loudspeaker dedicated to representing low frequencies, typically under 120 Hz.

USB - Universal Serial Bus, a type of external data interface for computers capable of up to 480 MBs operation. Similar to FireWire but with a generally "slower" data transfer rate and therefore it is better suited for lower audio channel counts. Most USB-capable audio interfaces feature less than 10 input and output channels.

Velocity - A MIDI parameter that defines how soft or hard a note is "hit."

To learn more about these terms, try search engines or online dictionaries (Type <term> define in Google for definitions) or wikipedia.org. On PCs, open the Help file for your recording software; for Macs, choose 'Help.'

Appendix C Recording Process and Gear List

What follows is my **recording process** and the **gear list** for my records *Whipped, Pagoda,* and *Penultimate,* with just a few minor differences in the process for *Penultimate.* For that record,

- Recorded drums *and* vocals at the engineer's (Ron Guensche) project studio, taking those tracks home as I did in the past with drum tracks.
- Ron recorded the drummer and the bassist playing "live" together to the scratch track.
- I recorded and mixed in Logic Express (I recorded and mixed in Cubase VST 5 for *Whipped* and *Pagoda*).

You can hear samples of all songs from all three of these records at www.MySpace.com/richardturgeon.

Here's the process:

1. **I built reference tracks** on my computer at home. In both Cubase and Logic Express, this simply entails setting a click track by choosing a BPM (or beats per minute) setting for each song, which constitutes any song's tempo. On top of that click track, I'd usually lay down a one-take, scratch guitar and vocal track to build on. As we covered previously, you can build any song on top of a click track and at least a scratch guitar track for reference.
2. **I recorded drums** in Ron's project studio with six mics on the kit. Why did I "outsource" the recording of the drums when I could just use my 4-input Tascam to record 4 channels: 2 overheads, a snare, and a kick?

   As a drummer myself, I like the additional mixing precision and control—and bigger sound—I get with 2 additional mics on the toms (rack and floor). I also don't have Ron's superb mic collection (expensive) and vast knowledge of mic placement and room acoustics. Ron also has an audio input unit with multiple inputs (expensive) and a superb engineer (Ron) to choose and place them properly and man the console. Finally, Ron also has the room for drums (Mine are in a rehearsal space and I can't play them at home because of the noise).

After we recorded my drum tracks, Ron burned the tracks onto a CD-ROM so I could take them back to my home studio.

3. Back at my home studio, I loaded the drum files into my audio software and made sure they were in sync with the click track. I usually like to record **rhythm guitars** next (I record through a Line 6 Pod so I can go direct and get a nice guitar sound without disturbing the neighbors with an amp), then lead guitars, then bass (I tend to record bass direct then just apply a software amp modeler preset, which can sound quite good).
4. I took my laptop, input unit, mic stand, mic and pop-guard to a rented rehearsal spot to record **vocals**. It's tough to engineer yourself if you're the singer, but not impossible (which is why I eventually chose to have Ron engineer vocals on *Penultimate*). If I'm engineering myself—as I did for *Whipped* and *Pagoda*—I record multiple vocal takes on multiple tracks (lead, then background vocal harmonies), then go through the vocal lines phrase by phrase, snipping out the ones I didn't want and keeping the ones I wanted on the screen (a process known as *comping*). Then you can just bounce down your lead vocal track from all the good stuff. I ultimately found it much easier to have Ron engineer vocals for *Penultimate* as it saved time on me choosing and editing takes that worked back home since he and I did a lot of that together as we recorded.
5. **Mixdown.** This starts when you're finished the recording phase of any project and get into the post-production phase. I never worked on a real console, instead mixing all my tracks strictly with my mouse and on my laptop screen. Not necessarily the best way to work, but certainly the most direct. No fancy hardware consoles or trackballs required to get the job done, just a good set of ears and a working knowledge of the software at your disposal.

In mixdown, I almost always do the drum mix first, then adjust the guitars, then the vocals last. I usually set the kick drum, snare, and vocal levels first, since they'll usually be out front in a rock record, and you can adjust everything relative to those.

Once I'm happy with a mix, I usually burn a CD and bounce down to an MP3 so I can listen on lots of different systems, including a boom box, the car, a portable CD player, and of course, my iPod, on different headphones for good measure. I walk around and outside the room to listen for little imbalances and nuances.

Take your time before deciding that your mix is final. Lay off it for a few days before these tests if you need to freshen your ears first. Not incidentally, I mixed *Whipped* while monitoring on a CD boom box that cost around $125, and *Pagoda* and *Penultimate* on JBL Creature Speakers. Again, not necessarily the best way to go (I'll at least get a decent set of monitors for the next record, but you do what you gotta do...). That said, don't wait for fancy gear to get your records done—just get them done and upgrade to better gear when you can, and only if you feel you need to.

GEAR

This list isn't to promote this particular gear or to put undue emphasis on gear—it's to demonstrate that one can make a professional-sounding record for not a lot of money.

- *Audio cables*—2 quarter-inch guitar cables and 1 XLR mic cable; all three are Monster Cable.
- *Two hard drives*—Both 7200 RPM Porsche LaCie drives (160 and 250 GB), both relatively lightweight. One is the main one I record to, the other my backup.
- *Line 6 Pod*—This hardware amp modeler allows me to record great-sounding guitar parts—direct and at home, with headphones on—without disturbing the neighbors. Tons of "amps" and common guitar effects to choose from in one lightweight piece of hardware. Great to practice with, too, of course.
- *Tascam US-428*—4-channel, USB-interface audio unit with 4 XLR inputs and a mini-mixing "console" I never use; 4 mic pres but no phantom power, and two are unbalanced. It only goes up to a 44KHz *sample rate* (as opposed to 96KHz or 192KHz; see Appendix for definition), but it does the job. Where this unit came in handy for me (once) was when I miced a drummer's kit with 4 Shure SM57 mics for a demo version of a song. Not as crisp or controlled as having 6 higher-end mics on a kit, but it sounded surprisingly good—thick and fat. I'd do it again in a pinch if I had to. I should note that I recently purchased an M-Audio Fast Track for a lighter, more portable input unit for those sessions where I only need one 1/4" cable or XLR (mic) input for vocals, guitar, bass, MIDI keyboard—i.e., pretty much everything except drums, where you'll typically want to record multiple sources at once (snare, toms, kick drum, etc.)
- *Logic Express*—Apple's music creation software, what I record into and mix in. Great plug-ins, great control, top-tier MIDI functionality if and when you start going beyond recording live performance tracks.

- *iBook G4*—You don't need a ton of computing power to make rock records. On the average, my songs tend to come in around 15–30 tracks, including 6 drum tracks. I produced my first LP, *Whipped,* on a first-model clamshell iBook and Cubase VST 5; on higher track counts, I did have to bounce a few rounds (drums first, of course, then guitars, then I'd do lead vocals last). But Logic's Freeze function, coupled with the 700+K of RAM I loaded into my 2005-model iBook, lets me pile up that 24-bit, 44.1KHz track count without worrying too much about running out of resources.
- *Headphones*—I literally mixed *Pagoda* on a pair of inexpensive Sony noise-canceling headphones and desktop LBJ Creature Speakers. I've already admitted that this isn't ideal and it's always better to have a higher-end pair of monitors to mix and master on (*See "Monitors" passage in Part 2*).

Index

R

S

T

V

Home Studio Setup

Everything You Need to Know from Equipment to Acoustics

Ben Harris

Home Studio Setup

Everything You Need to Know from Equipment to Acoustics

Ben Harris

AMSTERDAM • BOSTON • HEIDELBERG • LONDON • NEW YORK
OXFORD • PARIS • SAN DIEGO • SAN FRANCISCO
SINGAPORE • SYDNEY • TOKYO

Focal Press is an imprint of Elsevier

Contents

CONTENTS

CONTENTS

CHAPTER 3

Construction

Acoustic treatment is not something that you can completely buy in a store and have it work perfectly. There is always some construction involved in making acoustic treatment work with your room. Most articles and books about acoustics drive you one of two ways: All acoustic treatment can be built by you, or all acoustic treatment must be purchased from brand-name acoustic manufacturers. I feel that the answer is often a combination of these two polarities.

BASIC CONSTRUCTION TIPS AND TRICKS

There are many times where you could easily build something that will work wonderfully for your specific acoustic problem, and there are other circumstances where something premanufactured, built to scientific specifications, or cut with lasers would be the most effective solution. In this chapter we discuss some common construction tips for homemade acoustic treatment materials as well as the strengths and weaknesses of available products in the marketplace. Keep up to date by consulting the companion website, theDAWstudio.com, for additional techniques, product reviews, and new technical advancements.

Bass Traps

First things first: A couch does not function as a bass trap. Yes, it does accomplish some absorption, and because it is big it does absorb lower frequencies than a thin piece of foam, but it is not a bass trap. A *bass trap* is a device that significantly absorbs low-frequency energy. This is usually accomplished by having sound pass through a permeable surface, move into a space of air, and then not return into the original space.

There are many bass traps that you can purchase that are very effective. Most of these fit into corners because these are places where

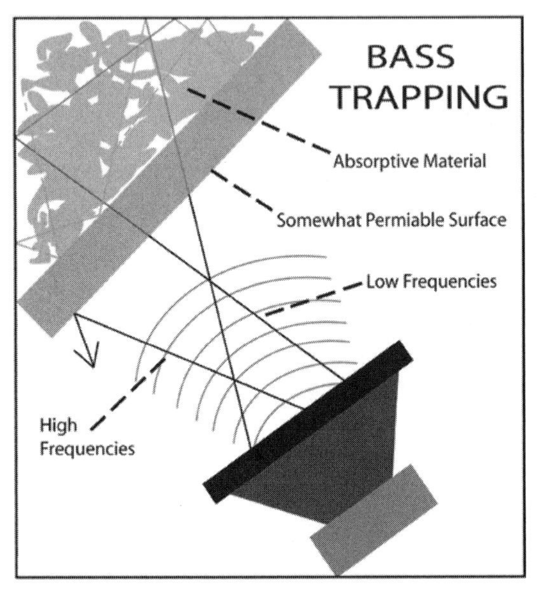

Figure 3.1
Bass traps let lower frequencies pass through a permeable surface, where the waves are then trapped and absorbed. Many bass traps reflect higher frequencies while letting low frequencies pass through. (Created by Ben Harris.)

Figure 3.2
A bass trap can be created by placing an acoustic panel diagonal in a corner. (Photo courtesy of Wind Over the Earth, Ben Harris.)

bass tends to build up (or load) and become problematic. There are also panel or membrane absorbers, which are considered bass traps, but use a different technique for trapping low frequencies. We'll talk about these later. The funny thing is that when it comes to traditional bass trapping (as mentioned first), custom-made devices and structures are usually the most effective and commonly used in professional facilities.

Building a Bass Trap

A bass trap can be made of a thin permeable wall with a space loosely filled with absorption behind it. Another type of bass trap, called a *circle trap* or *tube trap,* is basically chicken wire tied into a cylindrical shape covered in absorptive material and fabric with a stiff cover on the top and bottom.

The space in the middle is left empty to create a pocket of air to trap low frequencies. There are many alterations to this design using large PVC pipe with holes to pegboard covered in plastic. This type of trap has very simple construction but is often sold premanufactured at outlandish prices. The material and diameter of the tube have a direct relationship to which frequencies pass through, which frequencies are reflected, and which frequencies get trapped.

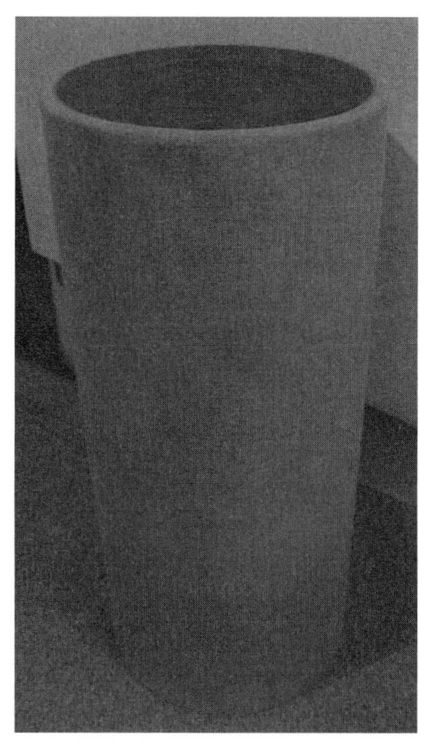

Figure 3.3
This is an example of a cylinder bass trap. (Photo courtesy of Wind Over the Earth, Ben Harris.)

A bass trap does not necessarily have to be a freestanding structure or something attached to a wall; it can be incorporated into the wall. You can build a second permeable wall next to an existing wall, cover it with fabric, and loosely fill the space with acoustic material such as insulation. The distance from the existing wall helps determine what frequencies are trapped, so it is a good idea to put the permeable wall on a slant, make it curved (convex), or make it stepped. This way it acts as broadband low-frequency absorption and trapping instead of focusing on a narrow frequency range. By making the outer surface somewhat hard and reflective while slanted, curved, or stepped, you can also create some diffuse surfaces.

Figure 3.4
A studio doesn't have to be in a square room. You can use slanted and stepped walls to help diffuse and scatter reflections. The walls are also an example of built-in bass trapping being built with stretched fabric over spaces filled with absorption. (Photo courtesy of Immersive Studios, Ben Harris.)

Diffusion

The main idea with diffusion is to scatter the acoustic energy back into the room to maintain the energy but not focus it back toward the listening position or a microphone. Three-dimensional diffusers are usually used above the listening position to avoid reflections off the ceiling from the speakers. These diffusers are used on the ceiling in the studio, especially over recording spots in the room, to avoid reflections from the source bouncing off the ceiling and hitting the microphone. Two-dimensional diffusers are usually placed behind the listening position and anywhere on the walls in the studio. Many people simply place diffusion behind the listening position because that is what everybody does, but it really doesn't work correctly unless there is a sufficient amount of diffusion in the back and enough reflection on the side walls to have that energy return to the listening position.

If that doesn't happen, the diffusion simply scatters the energy away from the listening position, which isn't bad, but it could be better. Diffusion in the studio is great when mixed intermittently with absorption. It doesn't work too well if there is only diffusion on one wall and absorption everywhere else. By placing diffusion, absorption, and reflection throughout the room, bad or unwanted reflections can be stopped and good reflections can be spread throughout the room to die out naturally. This will make the acoustics sound alive and natural.

You can use a three-dimensional diffuser on a wall, but energy that is diffused up and down is somewhat wasted. It is more than efficient

Figure 3.5
This image shows two-dimensional diffusion behind the listening position. (Photo courtesy of Immersive Studios, Ben Harris.)

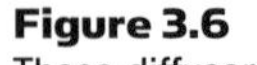

Figure 3.6
These diffusers scatter sound on a three-dimensional field. They are most commonly used on the ceiling above the mix position in a control room or the tracking area in a studio. (Photo courtesy of Sam McGuire.)

to diffuse energy to either side on a wall. That is why a two-dimensional diffuser works well on walls. Ceilings need to diffuse energy in every direction, which is why three-dimensional diffusers are more common on the ceiling.

Figure 3.7
A simple bookshelf can provide a decent amount of diffusion in a studio. (Photo courtesy of iStockphoto, Luoman, Image #7106618.)

Buying or Building Diffusers

Many companies make expensive and very effective diffusers. RPG makes a two-dimensional diffuser (usually used on a wall) and a three-dimensional skyline diffuser (usually used on a ceiling). These diffusers sound great just about anywhere you place them. They are mathematically designed to equally scatter a range of frequencies. The frequency range of diffusion capable of a structure directly correlates to the distance of the gaps, spaces, or depth of the material. For example, if two slats of wood in a diffuser are 2 inches apart from each other, the lowest frequency affected will be 3500 Hz. (Refer to the frequency chart in Chapter 1.)

You don't necessarily have to buy or build a specific product to achieve more diffusion in your room. Any structure that has hard surfaces at different layers can achieve some amount of diffusion. For example, a bookshelf is much better than a straight wall. Just think of how different a room sounds when it is empty, as opposed to when it is filled with furniture, bookshelves, and wall hangings. All these elements add varying levels of absorption and diffusion.

There are lots of examples of three-dimensional art pieces that have multifaceted layers that are perfect for diverting acoustic energy while making a room look inviting and artistic. Be creative with what you place in your room, always keeping in mind how it might affect the acoustics. Any curved wall, pillar, or structure will scatter acoustic energy as it reflects it. The key is to have a convex surface such as a pillar (which scatters the energy), not a concave surface such as a bowl (which focuses the energy).

Figure 3.8
Acoustic foam is only effective in absorbing high-mid to high frequencies. (Photo courtesy of iStockphoto, Clay Cartwright, Image #116644.)

The Problem with Foam

Quick-fix acoustic foam products have infiltrated the home studio market in the last few years. These products are wonderful to market and sell for a few reasons. First, they are easy and cheap to make; second, they are easy for the customer to install (requiring only some glue or staples), and third, they appear to be a quick fix to a complicated problem.

But the problem with foam is that it isn't always a quick fix. There are often better and cheaper solutions, and though foam might fix some problems, it can't fix all of them. What often happens is that someone buys some foam (1, 2, or even 3 inches thick), places it all over the walls, and then proceeds to record or mix. They find that the foam effectively absorbs the high-mid to high frequencies, but it really doesn't do much for low-mids and lows (the real problem frequencies in a small room). It also fixes issues with flutter and ringiness. This is great, but now all the high end is gone and the low end is exactly the same as before, without any highs to balance it out. Now the room sounds insanely muddy and boomy, with no air or life to it at all. The problem with foam is that it leads users into a false sense of security, just like 50 SPF sunscreen and low-fat snacks. Foam does not fix everything; it can help, but it is not the end-all solution.

Figure 3.9
Fiberglass insulation such as this can be wrapped and covered with fabric or placed in bass traps to effectively absorb a wide range of frequencies. (Photo courtesy of iStockphoto, Branko Miokovic, Image #2534690.)

Alternatives to Foam

If you go to a professional studio and check out their acoustics, nine times out of 10 you will not see foam anywhere in the studio. This is because there are a handful of alternatives that are usually cheaper and more effective. The first is fiberglass insulation. This is used to insulate buildings and houses from heat and sound transmission. Pieces of this insulation are usually placed on walls, in frames, or in bass trapping structures and covered with fabric to prevent airborne fibers. This material has commonly been used for years due to its availability and effectiveness. It even comes in different sizes and with varying ratings that can help you decide which frequencies will be more effectively stopped.

The problem with this material is the fibers. These fibers rub onto your skin, go airborne and into your lungs, and some say cause adverse health affects to the user. This is why many people use alternate forms of insulation. There are many health- and eco-friendly products available for building materials, and therefore acoustic materials as well. Recycled denim and cotton insulation is a great example. This is similarly priced to fiberglass insulation but much easier to work with and just as effective. Some acoustic companies sell this same generic product specifically as acoustic treatment at twice the price. Recycled cotton and denim insulation can be used in all the same ways as fiberglass insulation except for the fact that it does not have to be completely sealed from contact with the air you breathe. This gives you limitless options as to how you utilize this insulation for acoustic treatment.

In addition, other types of insulation such as shredded newspaper, blankets, and carpeted wall coverings can serve as alternatives to foam for many acoustic absorption needs.

Building Additional Permeable Walls

By understanding the three types of acoustic treatment—absorption, diffusion, and bass trapping—and understanding how each is accomplished, you can easily build acoustic treatment that is multifunctional, ergonomic, and visually pleasing. Building additional permeable walls is an effective way of combining the three types of acoustic treatment techniques to effectively combat acoustic problems. Basically, building an additional permeable wall is building a large bass trap. By having a wall that sound can pass through with a space filled with absorptive material behind it, you can effectively trap a lot of troublesome bass frequencies. If you then curve, angle, or step that wall, you not only add diffusion, you vary the distance of the gap behind the wall. This variation

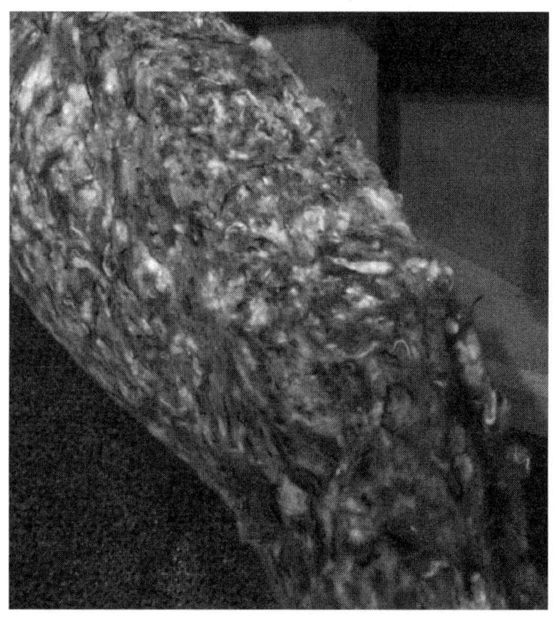

Figure 3.10
Insulation made from recycled cotton and denim has similar properties to fiberglass insulation and is much easier to work with. (Photo courtesy of Wind Over the Earth, Ben Harris.)

directly corresponds to the frequencies that will be trapped. (See the frequency chart in Chapter 1.) Now your bass trap becomes more effective in a wider range of frequencies.

Acoustic ceiling tile, carpet, and pegboard are a few examples of material that can be used to create these permeable walls. The main thing to remember is that you are trying to create a balance of absorption, diffusion, and bass trapping. Cater your structure to the needs of the room and be prepared to change and rebuild if it doesn't sound right.

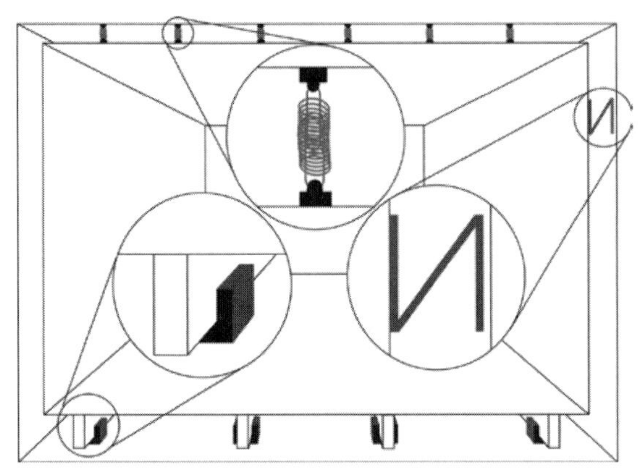

Figure 3.11
Here are three different products used in construction to help decouple walls from each other. These products are examples of many different techniques used to assist in acoustic isolation. (Created by Ben Harris.)

Wall and Floor Construction

While discussing isolation, the first chapter put a large focus on "mass … decoupling … mass," which essentially is building two structures while not letting them connect to each other. That's great; so we'll build a floor, put a layer of air, and another floor resting on the air. But … that's impossible? It is, and that's why there are many ways to essentially decouple structures while still obeying the laws of gravity.

There are some great devices and materials that make this possible, including springs, rubber, and foam. All these products help absorb vibrations, which is essentially decoupling the mass from the next mass or structure. If the two structures are connected, vibrations pass directly from one to the other. If there is a space of air between the structures, the vibrations will be trapped or greatly decreased because there is no mass to pass them along. If there is a pocket of air and then stiff brackets connecting the two masses to each other, the vibrations will pass through the brackets. The solution is to use materials such as springs, rubber, and foam to connect the structures but not rigidly couple (or connect) them to each other.

One way to decouple a raised floor from the main floor is to build the raised floor on studs resting on rubber pads. These pads will help decouple the raised floor from the main floor. Another technique is to use foam panels between the floors. This accomplishes the same thing as the rubber pads. Two walls (or ceilings) built next to each other can be decoupled by connecting them to each other with springs or a springlike Z-shaped bracket. These brackets will absorb vibrations and minimize them from passing from one wall to the other. If you have already effectively built a raised floor, build the walls on that floor and the ceiling on those walls so that the only connection that separates the room from the larger room are the rubber pads that the floor is sitting on. In larger studio design you can get even more crazy, creating two separate foundations separated by only dirt and then building each wall on an individual foundation. The vibrations stop when they hit the dirt and do not pass through to the other wall.

Check the DAWstudio.com for more details on acoustic products used for isolation in wall and floor construction.

Figure 3.12
A side view of dual window and wall construction utilizing varying glass thickness, one slanted piece, and air between for decoupling. (Created by Ben Harris.)

There are a few additional problems to avoid when placing windows in these walls so that you can see your clients. If both pieces of glass are parallel and the same thickness, they will vibrate in synchronization with each other, passing the vibrations right through to the other side. The solution is to use panes of glass of varying thicknesses so that they don't resonate at similar frequencies, then place one window on an angle to vary the distance between the two pieces, further preventing passing vibrations. The other big problem with windows (and doors) is that they are basically large holes and if air can leak through, so can sound. The solution is to seal all windows and doors as though you were sealing external windows and doors from the heat or cold outside. All these techniques will help isolate one room from another.

Panel or Membrane Absorber

Panel and membrane absorbers are a great solution to dealing with bass problems in situations with space limitations. These absorbers take up less space than traditional bass trapping because they function on a different acoustical premise. Instead of trapping the energy with absorptive materials and a pocket of air, panel and membrane absorbers utilize a thin layer of wood or heavy fabric that will vibrate but then dampen that vibration. Panel absorbers usually have a

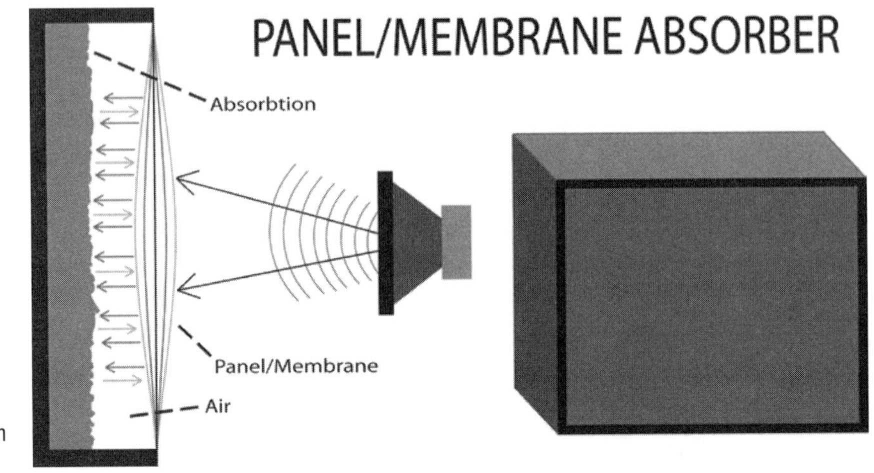

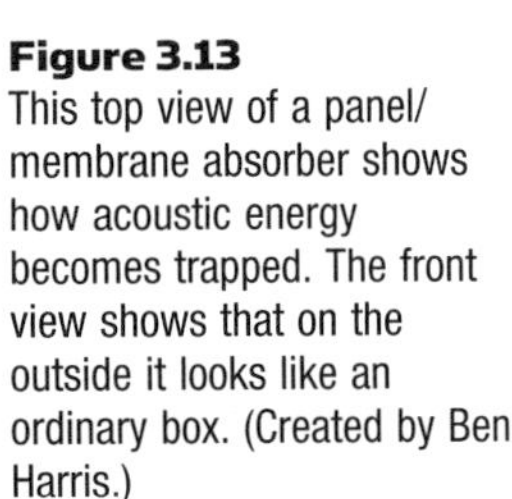

Figure 3.13
This top view of a panel/membrane absorber shows how acoustic energy becomes trapped. The front view shows that on the outside it looks like an ordinary box. (Created by Ben Harris.)

wood panel functioning as the absorber, and membrane absorbers can be built from roofing paper, synthetic leather, or any heavy tarp-like material. The idea is that the air behind the panel or membrane is sealed and contains foam or insulation (not touching the panel or membrane) so that when the front of the absorber begins to vibrate, the backside cannot keep up with the vibrations of the front. This is because the air on the backside is sealed and cannot move freely. As a result, the acoustic energy that excited the front of the panel or membrane becomes absorbed as heat into the air behind the panel and therefore does not return into the room.

These absorbers can easily be built with materials from a hardware store. A panel or membrane absorber is basically a shallow box with a thin layer of foam or insulation inside, pressed toward the back-side. There is a layer of air, and then the front is covered with either a panel or membrane made of the materials mentioned earlier. The depth of the box is in direct relation to the frequencies to be targeted. Frequencies that measure two and four times the length of the depth of the cabinet will be targeted the most by the absorber, but the material used for the panel or membrane is also a big factor to which frequencies are affected. The difference between using a wood panel or a membrane mainly affects the frequencies being targeted. Wood panels usually function more effectively on midrange and high frequencies due to their rigidity, whereas membranes usually function better on low-mid frequencies.

COMMON PROBLEMS AND SOLUTIONS

We've talked about a lot of common acoustic problems through the last few chapters, but the idea for this section is to address the issues not covered so far as well as revisit some covered earlier in greater detail. Some of the issues include room modes, equalization, and overabsorption.

Room Modes

Room modes are the frequencies that become problematic in a specific space based on the three main dimensions of a room; height, width, and length. Any one dimension in a room will produce problematic frequencies based on that dimension's length. For example, if the length of a room is 11 feet, then 100 Hz (one wavelength equal to the room length) will bounce back and forth on top of itself, effectively canceling itself out; 150 Hz (one-and-one-half wavelength) will pile on top of itself and double in amplitude; 200 Hz (two wavelengths) will do the same thing as 100 Hz.

This is problematic, but it becomes worse when we add two other dimensions to the room (height and width). What if your room is a perfect cube? Then each dimension has the same problematic frequencies, making the room an acoustic nightmare. Having one dimension twice the length of another is also problematic from the same reason as the cube. The way to calculate your room's modes is

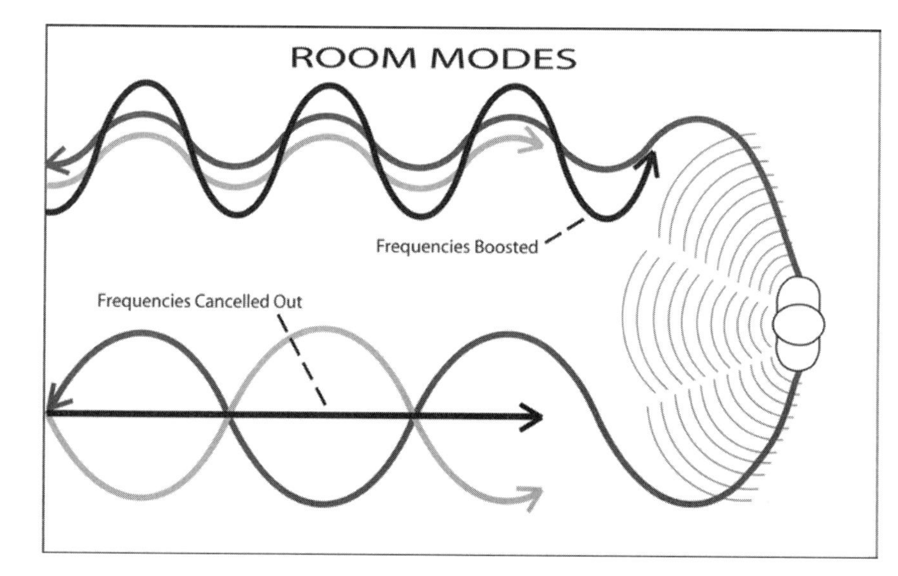

Figure 3.14
Room modes occur with the frequencies that have similar full, half, double, or any fraction wavelengths of a room's dimensions. As shown, frequencies will either cancel themselves out or boost themselves, depending on what point of the cycle a wave is at when hitting a boundary and reflecting. (Created by Ben Harris.)

to make a list of each dimension of your room and all the frequencies that have wavelengths that are multiples of your room's dimension (half the length, the same length, one-and-one-half the length, twice the length, and twice each of those). Once you do this for all three dimensions, find the frequencies that share two of the three dimensions, and those are your room modes. If there are common frequencies on all three lists, they will be specifically problematic. These are the frequencies that you should specifically target in you acoustic treatment efforts. Don't go too crazy; just focus on those specific frequencies. You want a balance of treatment with diffusion, absorption, broadband low-frequency absorption, and then specific treatment focusing on the problematic frequencies.

Equalization

Equalization is often misused in acoustics when it is used as the first line of defense. Often people will analyze a room, see the dips and bumps in the analysis, and go crazy with an equalizer trying to correct the problems. The problem with this lies in the previous paragraph. If you understand what is causing the frequency variations, you realize that equalization isn't going to fix the problem and will often make it worse. Think about it this way: If a room has a dip at 200 Hz because that is a frequency that cancels itself out as it bounces along the length of the room, adding more information at 200 Hz to the room will actually cancel itself out more and create a bigger dip. The same thing goes for bumps caused by frequencies bouncing on top of themselves in phase. There are also other factors that cause frequency deviation, such as reflections from the back of the speakers off the front wall that cause comb filtering, reflections off the side walls doing the same, and resonating furniture and items in the room. The problem is that trying to solve all these complex problems with a little equalization is simply illogical.

The logical approach to using equalization is to not use it as the first line of defense. First use techniques discussed in these chapters to correct these complex problems occurring in a room. Then, when you've done all that you can with physical acoustic treatment and it still isn't as flat as you would like, add a little equalization to balance it out. Don't just throw on a cheap three-band or anything. If you are going to equalize your room this way, you have to use a high-quality graphic or parametric equalizer. The adjustments are going to be very small—probably no greater than one or two dB of gain or attenuation (turning the signal down).

Absorption

If you go to a music store and tell the salesperson that you need some acoustic treatment for your studio, what will he suggest? Foam! But foam is a small and inadequate solution to a big and complex problem. The main reason that foam is readily available at music and recording equipment retailers is because it is an acoustic solution that can easily be packaged and sold. The problem is that it is not the end-all solution.

What happens is that people notice that their home-recording space sounds ringy (flutter echo) and boomy (low-frequency room modes). They buy a bunch of foam, glue it on their walls, and now the ringiness is gone but the boominess is worse. This occurs because the foam does not do much for the low and low-mid frequencies. It is great at getting rid of flutter echo and dampening the high and high-mid frequencies, but that doesn't do anything to solve the boominess. In fact, the boominess now seems worse because there isn't as much high-frequency information to counterbalance it.

What is the solution when you go back to the music store and explain the problem? Buy bass traps. Now you drop even more money on foam and glue these up in the corners of the room. Now it sounds a little better. The low-mids are not as horrible, but the low frequencies are as bad as ever and you still have no high frequencies in your room.

This process could go on forever, but the answer is that there is no kit or perfect solution that you can buy at a store, glue to your wall, and solve all your acoustic problems. Each room is unique and requires different treatment, care, and considerations. By understanding acoustic fundamentals, room interaction, and treatment techniques, you can individually treat a unique room to create an excellent recording and mixing environment.

Keep up to date with additional building tips and tricks, new building materials, and more detailed solutions to common acoustic problems at theDAWstudio.com.

Secrets of Recording

Professional Tips, Tools & Techniques

rne Bregitzer

Secrets of Recording: Professional Tips, Tools & Techniques

Lorne Bregitzer

AMSTERDAM • BOSTON • HEIDELBERG • LONDON
NEW YORK • OXFORD • PARIS • SAN DIEGO
SAN FRANCISCO • SINGAPORE • SYDNEY • TOKYO
Focal Press is an imprint of Elsevier

Contents

CHAPTER 5 • Emulated Effects

CHAPTER 6 • Adding MIDI Tracks to Recordings

CHAPTER 7 • Mixing Techniques

CHAPTER 8 • Mastering the Recording

CHAPTER 2
Timing Correction

INTRODUCTION TO TIMING CORRECTION

Correcting the timing of a piece of music, whether it be the drums, percussion, or merely tightening up doubled vocal parts, is one of the most important things that can be done to improve the quality of a recording. If faced with the choice of having a track be out of pitch or out of time in a recording, it is less appealing to the ear to have something out of time. Having parts that are out of time denotes a sloppiness to the performance, which translates to a sloppy recording. The ear is generally more forgiving to pitch, within reason, than timing.

In the world of editing tracks visually, it becomes easy to edit the timing so that tracks "look right." It is important to realize that there is musicality to the slight fluctuations of timing as an artistic license. Some parts sound better slightly behind the beat. When professional musicians are playing together, they know which side of the beat to be on for the style of music and their particular instrument. This is another instance where studying the style of music that you are recording will make you a better engineer.

There are two main types of timing correction. The most common method creates edits in the audio regions and moves them so that the transients are then on the appropriate beats. This can be done either manually or through software features that automate this process, such as Digidesign's Beat Detective.

The other main type of timing correction will actually stretch or compress the time of regions to make them fit the tempo of the song. This can create problems with artifacts created by the digital processing that is used to expand or contract the pitch.

Tools for timing correction

Most timing-correction software tools are proprietary to their host digital audio workstation (DAW) software. Most plug-ins used in this book are available across multiple platforms, as they function as an insert across a track. With timing correction, it is more a feature of the workstation software, as most of the handling of the audio does not take place in real time; they are merely editing features. This chapter will give you an overview of the main features used in this book.

Time compression and expansion

Depending on the algorithms used by a particular time-compression and -expansion feature, there may be artifacts created by the process. This is particularly more noticeable in time expansion as opposed to compression.

Percussive sounds have a specific transient that denotes the beginning of the sound. When these percussive sounds are stretched, oftentimes the transient can be smeared or even replicated to create the lengthened sound. A short time expansion can create a smeared transient. If the transient is smeared, the percussive nature of the sound can be diminished. With even longer time expansion, the transient can be replicated, creating a flam in the sound.

In Figure 2.1, there are five versions of the same audio file, which is a snare drum. It has a solid transient but a moderate decay to the sound. The upper track is the original track. The second version is time expanded by 100 percent, where you can see a double transient created. The third version of the snare drum is time expanded by 50 percent, where you can see a smearing of the transient. Take note

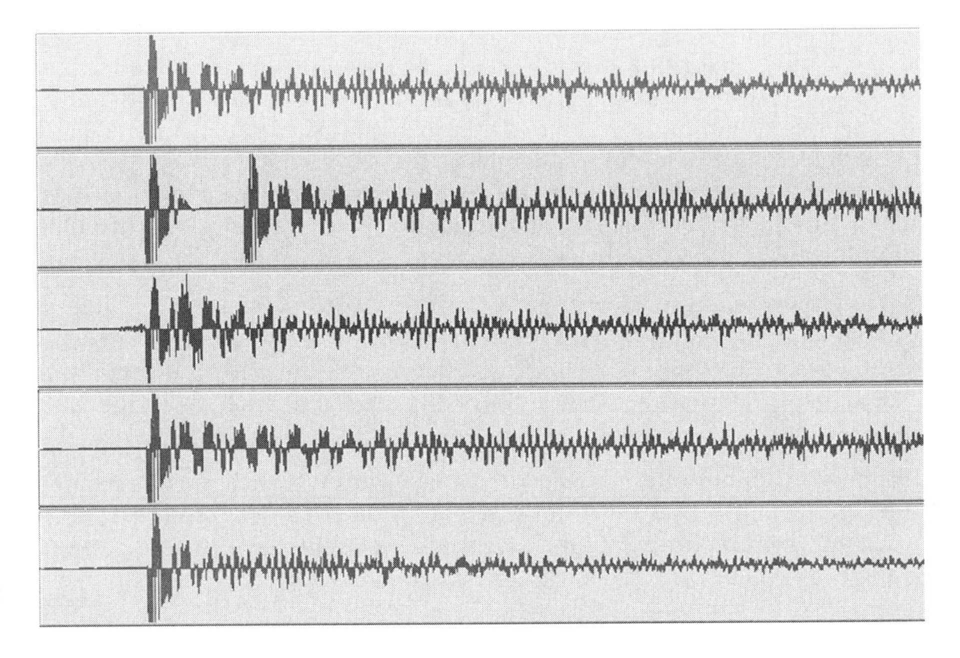

FIGURE 2.1
Five tracks, displaying the effect of time compression and expansion with different algorithms.

that this will create problems in the sound, as it will not cut through the mix as a snare drum should. The fourth version has the snare drum expanded by 100 percent, but with an algorithm that is specifically designed for percussive sounds. This will maintain the transient of the sound but increase the decay. The fifth version of the snare drum is time compressed by 50 percent, and you can see little effect to the transient.

Manual timing correction

When correcting the timing of a performance, there are a couple of different ways that various DAWs can handle the job. The first method that you will usually encounter is manually correcting the timing of a performance. This works well for subtle mistakes, such as the bass player hitting a note slightly earlier. A simple edit and nudging of the audio region will take care of this.

In Figure 2.2, you can see that the second note of the lower track, which is the bass, is behind the beat when compared to the kick drum hit above it. This is a classic example when a simple region nudge edit will correct this issue.

In Figure 2.3, you can see that the region of that note is separated from the surrounding notes, allowing for the region to be nudged to its appropriate position.

The region has been nudged, as seen in Figure 2.4; however, there is a gap between the end of the note and where the next note begins. This is where you will need to make a decision as to how to correct the empty space. In some

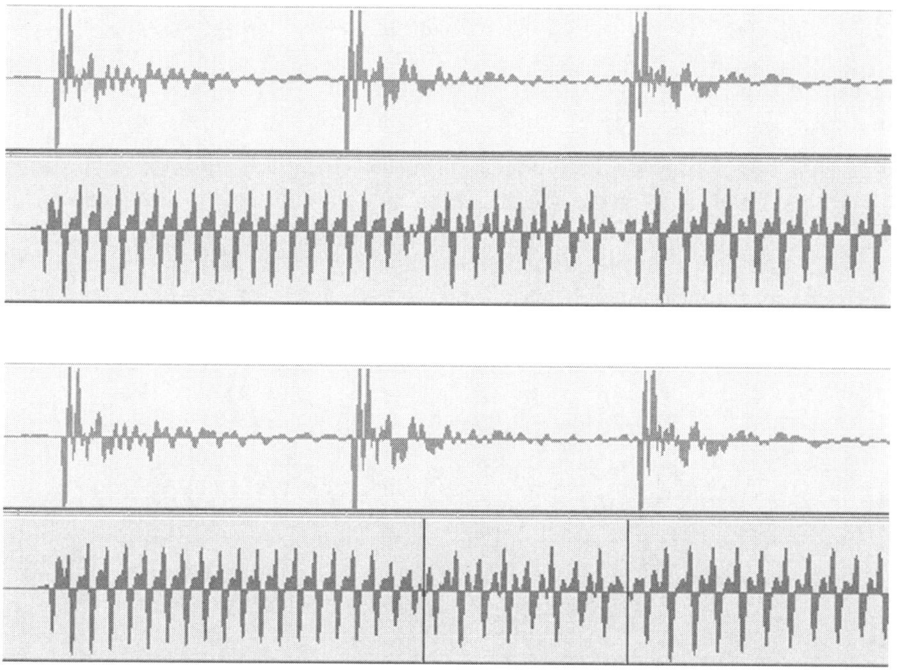

FIGURE 2.2
Kick drum (top), and bass line with the second note out of time when compared to the kick drum (bottom).

FIGURE 2.3
The region in the bass track has been separated in order to shift the note.

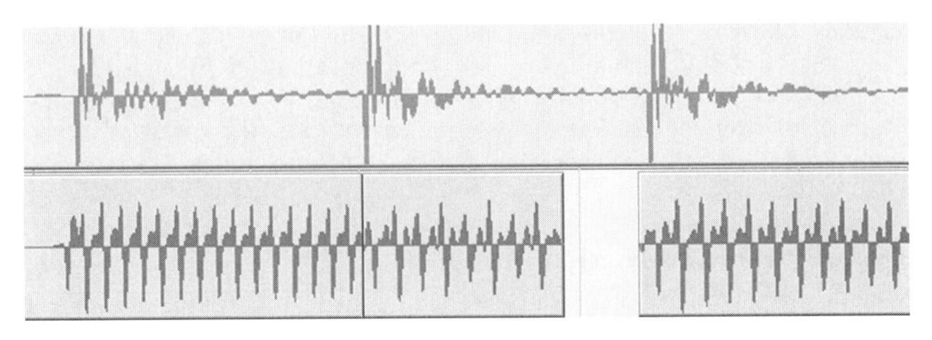

FIGURE 2.4
The region of the bass note has been nudged to be in time with the kick drum.

situations, the way that the edit is nudged, without any region adjustment, will sound just fine among all of the other tracks. The best way to make that determination is to listen to all of the other tracks playing at the same time to determine if the gap is noticeable or not.

Sometimes it is best to err on the side of caution and to correct the empty space. This will cover the edit, in the event that the song undergoes some rearranging in the future that would make the empty gap more apparent to the listener.

The first method of correcting this gap is to appropriately drag the end of the note to the right of the edit, back into the edited region. This edit will only work if the edited region has sufficient duration so that the dragged region can overlap during the sustained portion of the region. This is a good place to keep in mind the principles of utilizing zero-crossings for the edit. You may have to slightly nudge the timing of the initial region to create an overlap of the edited region in order to create the appropriate zero-crossing of the edit. In Figure 2.5, you will see that the third note has had its beginning region trimmed into the sustain of the corrected region, creating a longer, sustaining note.

Once the final edit has been made, you can determine whether or not you wish to crossfade the tracks to create a more transparent edit. This is not necessary to do in the beginning of the edited region, since it begins on an attack of the note, but in Figure 2.6, you will see that the edited region has had its sustain edited over the duration of one period of the waveform on either side.

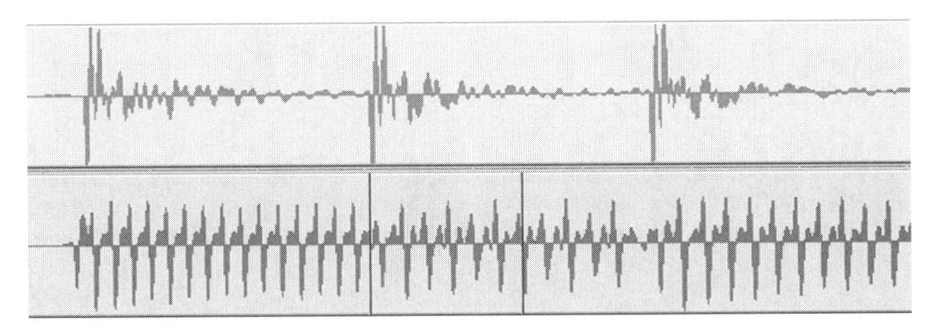

FIGURE 2.5
The corrected bass note has had the end region nudged earlier to correct the gap.

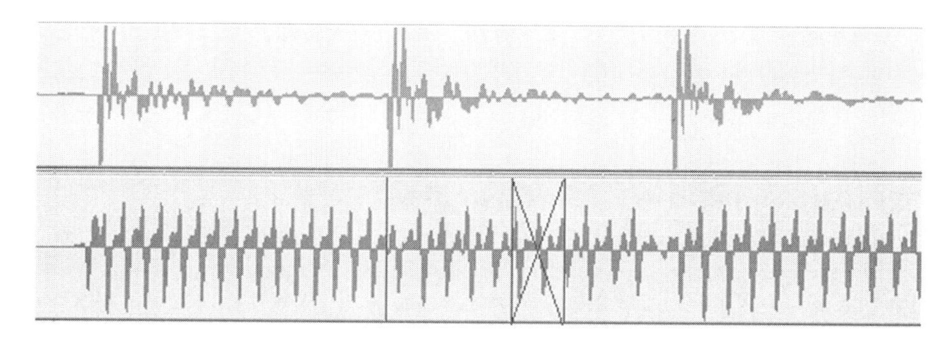

FIGURE 2.6
The nudged end region has been crossfaded to create a smoother edit.

BEAT DETECTIVE

One of the most powerful automated editing features added to Digidesign's Pro Tools in recent years is Beat Detective. Originally only available for the HD system, it is now available as a single-track feature in Pro Tools LE, or as the multitrack version found in HD. You can add the multitrack feature to Pro Tools LE with the purchase of the Music Production Toolkit, which also adds a whole host of other features, such as SoundReplacer and the ability to have a session with 48 audio tracks.

Beat Detective is a powerful editing tool that is used to tighten up rhythmic tracks according to the tempo map in Pro Tools. It can also be used to generate a tempo map from an existing performance.

Beat Detective accomplishes this editing by cutting up the regions according to where it detects the transients. It then moves the regions that it has created and places them in the correct time according to the tempo map in Pro Tools.

Steps of Beat Detective

When editing drum tracks with Beat Detective, there are basically three steps, with an optional fourth in the operation of the software to create the appropriate tracks.

- *Region separation*. Region separation will identify the beats performed by the drummer and separate them at the beginning of each selected transient.
- *Region conform*. Region conform then takes these separated regions and moves them to the appropriate bar, beat, or sub-beat in the measures.
- *Edit smoothing*. Edit smoothing can then move the ends of the regions to the adjacent edit so that there will not be any silence between some of the edits. In addition, you have the option of adding a crossfade at each of the edit points.
- *Consolidation of the tracks (optional)*.

At the end of each of these steps, you need to listen to the tracks being edited in order to check for accuracy. Beat Detective is not a completely automated process, but a user-guided process. Careful listening will help to avoid artifacts that may crop up due to misdiagnosed transients.

When you should use Beat Detective

Beat Detective can be used to generate a tempo map, as mentioned in Chapter 1, and it can also be used to put performances in time with an existing tempo map. Beat Detective is great for putting drums in time with the click track or locking a percussion track in time with the drums.

Beat Detective should be used right after basic tracking, before any of the overdubs have been recorded. If the other instruments are following the drum tracks, and you tighten the drum tracks to the tempo map of the song, by correcting the timing of the drums, the other instruments will sound out of time. Keep this in mind when planning your session so that you do not have parts that are out of time, in which case you will need to rerecord overdubs again.

Saving a new session

When doing any semidestructive editing, it is always good to save a new session file in your session folder. Also, it is a good habit to get into to save a new session file every time you do something major in Pro Tools, from editing to mixing. A helpful method would be to increment a number and add a note as to what is contained in the new session. If you are working on "Song A 1," save a new version as "Song A 2-Beat Detective." This will help you go back at any point in the future in case something gets messed up along the line.

Grouping the drum tracks

Beat Detective will work across multiple drum tracks if you have the multitrack Beat Detective with your software. In order for it to work across all of the drum tracks, they all need to be selected. The easiest way of doing this is to put them all in one group by selecting the names of all of the tracks, then selecting "Group ..." from the Track menu.

Duplicating the playlist

It is always good to create a safety net for yourself when doing semidestructive editing to your tracks. This can be accomplished by duplicating the playlist in Pro Tools so that you can always go back to an unedited version. It will be helpful to duplicate the playlist at the beginning and before starting each phase of using Beat Detective. This way, if you make a mistake, it is easy to copy and paste a previously unedited version on top of your active playlist.

To duplicate the playlist, make sure that the tracks you are using in Beat Detective are in a single group. Click on the triangle to the right of one of the track names and select "Duplicate Playlist." This will create a duplicate of the playlist so that you can now always go back to the original. The name of the playlist increases the name by .01, so keep track of what playlist you are working on. See Figure 2.7 for an example.

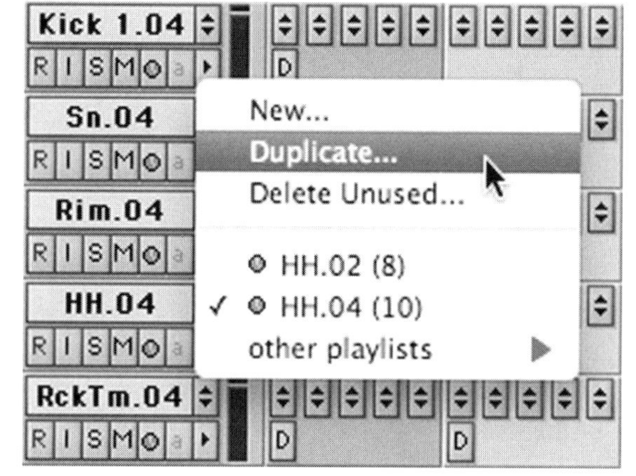

FIGURE 2.7
Duplicating the playlist in Pro Tools.

Creating sections to edit

In much the same way that we created a tempo map from the drummer's performance in Chapter 1, using four or eight bars at a time, we will use the same smaller sections to correct the timing of the drums here.

The reason to choose shorter selections is that Beat Detective gives you a wealth of options for resolution. You can adjust the resolution from a single bar down to 32nd notes. Since not every measure that the drummer plays will contain the same rhythmic resolution, it is best to break up these sections into the exact rhythm performed in the bar. A simple drum groove may only contain quarter notes or 8th notes, but when a drum fill is played, there can be 16th notes in the bars that contain the fills. Beat Detective will work best when given the most accurate rhythmic information according to the content of the track.

In the example in this book, the tempo of the session is 68 beats per minute. With a slower tempo, inexperienced drummers have a tendency to rush the beat, so the use of Beat Detective will be important in creating a solid and steady feel to the song. Because of the slow tempo, we will be selecting two measures to edit at a time.

Selecting sections works well by separating the two-bar region precisely. This is done by zooming in and placing the cursor right before the first hit of the kick drum as it hits right on the downbeat.

Selecting the region for editing

When you run the "Region Separation" phase of Beat Detective, it will automatically separate the region for you; however, it can be helpful to manually select the beginning and end points of the measures you are working with in small chunks to keep track of the small sections as you work through the session.

In Figure 2.8, you can see that the region has been manually separated at the beginning of bar 70, where the drums begin. The region can be manually separated by pressing "Command-E" (Mac) or "Control-E" (PC). If single-key shortcuts are enabled, simply pressing "B" will create the appropriate edit.

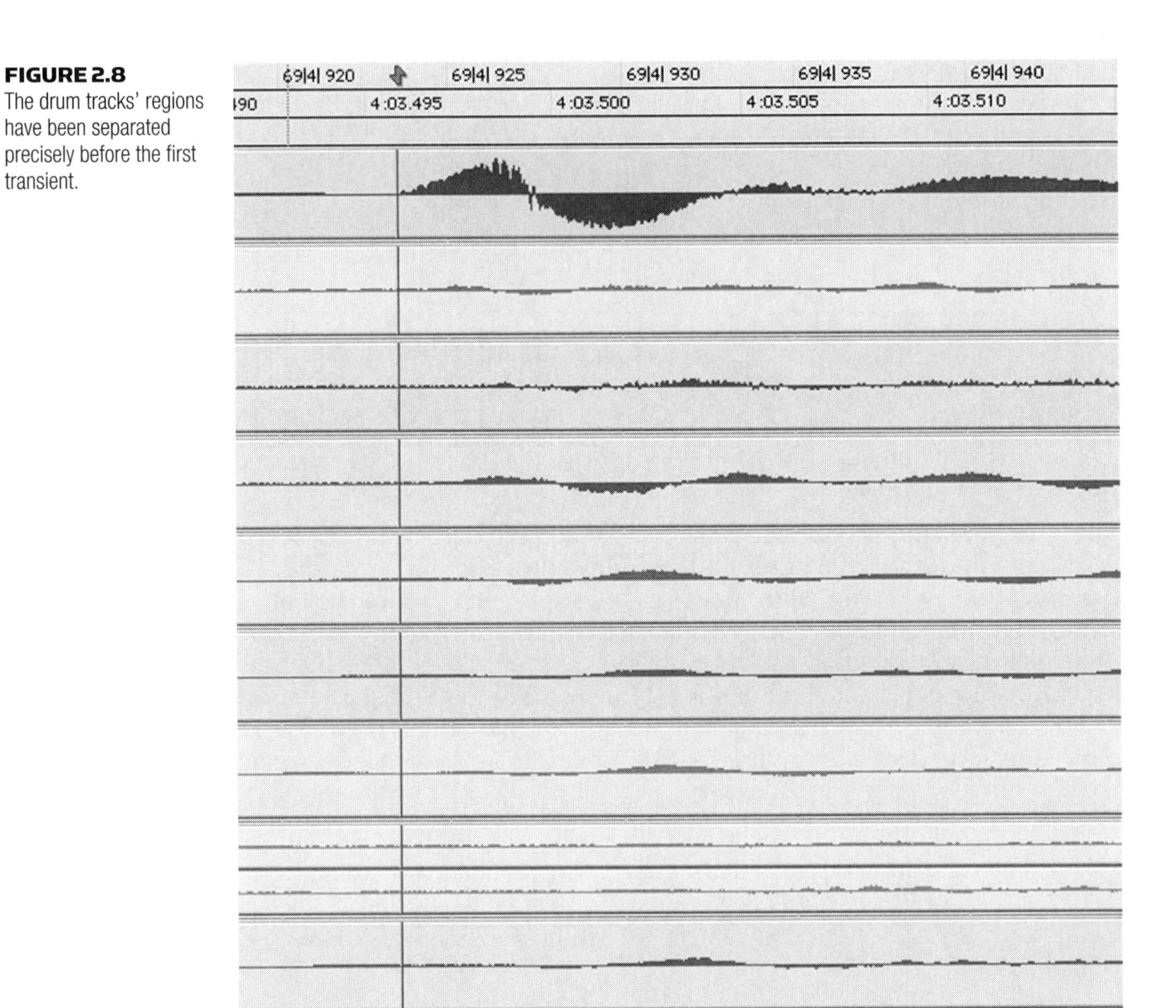

FIGURE 2.8
The drum tracks' regions have been separated precisely before the first transient.

The beginning of bar 72 is then separated in the same way so that we have the exact two bars selected that the drummer played. Now we want that section to conform in time to the same corresponding four bars in Pro Tools.

In Figure 2.9, you will see the region for the two bars, as performed by the drummer, separated but not edited yet. Make sure those regions are selected by double-clicking in the middle of the separated region. From this section, we can either manually enter in bars 70 and 72 for the Start Bar and End Bar, or we can take the highlighted tracks and click on "Capture Selection."

To begin the editing process, we will need to make sure that "Region Separation" is selected under "Operation" on the left (Figure 2.10). Upon listening to the track, determine the shortest rhythm that is contained in the performance. Beat Detective will be more accurate with the longest timing selection that is used in the performance.

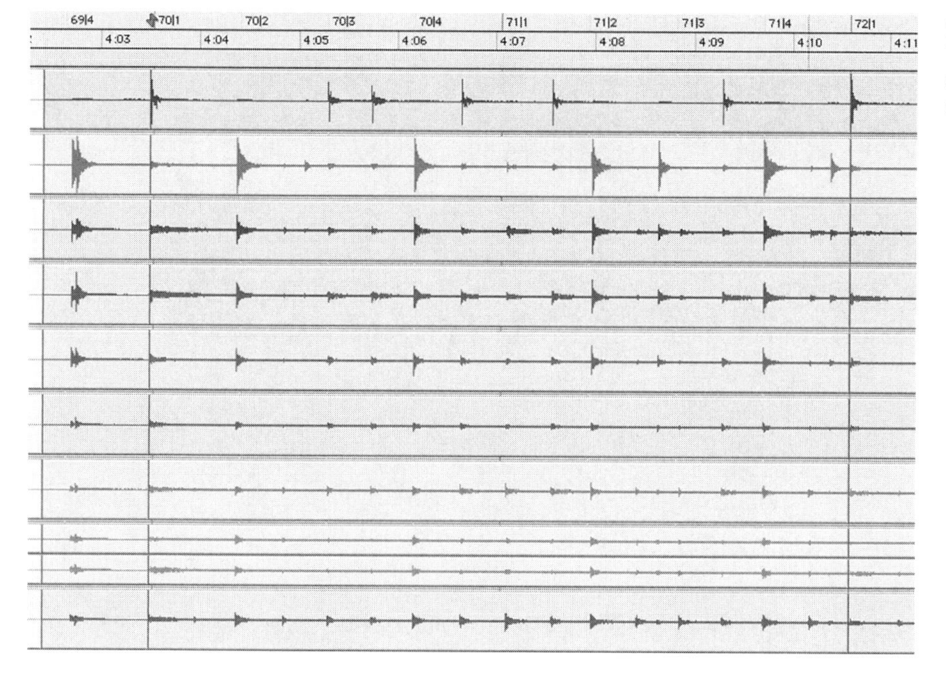

FIGURE 2.9
Two measures of the drum tracks have been separated.

Tabbing to transients

Pro Tools has a "tab to transients" feature that you may be tempted to use, as it will advance the cursor to the next transient that it detects. Even though this is given as an option, it is more accurate to do this manually, as "tab to transients" can place the cursor in the middle of the transient, and you will usually want to place your edit point right before the transient.

Selecting the appropriate resolution for selection, analysis, and separation

After you have the appropriate measures separated and selected, it is now time to get Beat Detective to automatically separate the regions (Figure 2.11).

In this selection you can see that the two bars contain 16th notes, so we will select "Contains" and set it to 16th notes. Any setting shorter than that can identify mistimed 16th notes as being 32nd notes. Additionally, if the selection contains 16th notes, and you have selected that it only contains 8th notes, then Beat Detective can possibly identify 16th note transients and place them on an 8th note beat when it comes time to conform.

FIGURE 2.10
Selecting the Region Separation operation in Beat Detective.

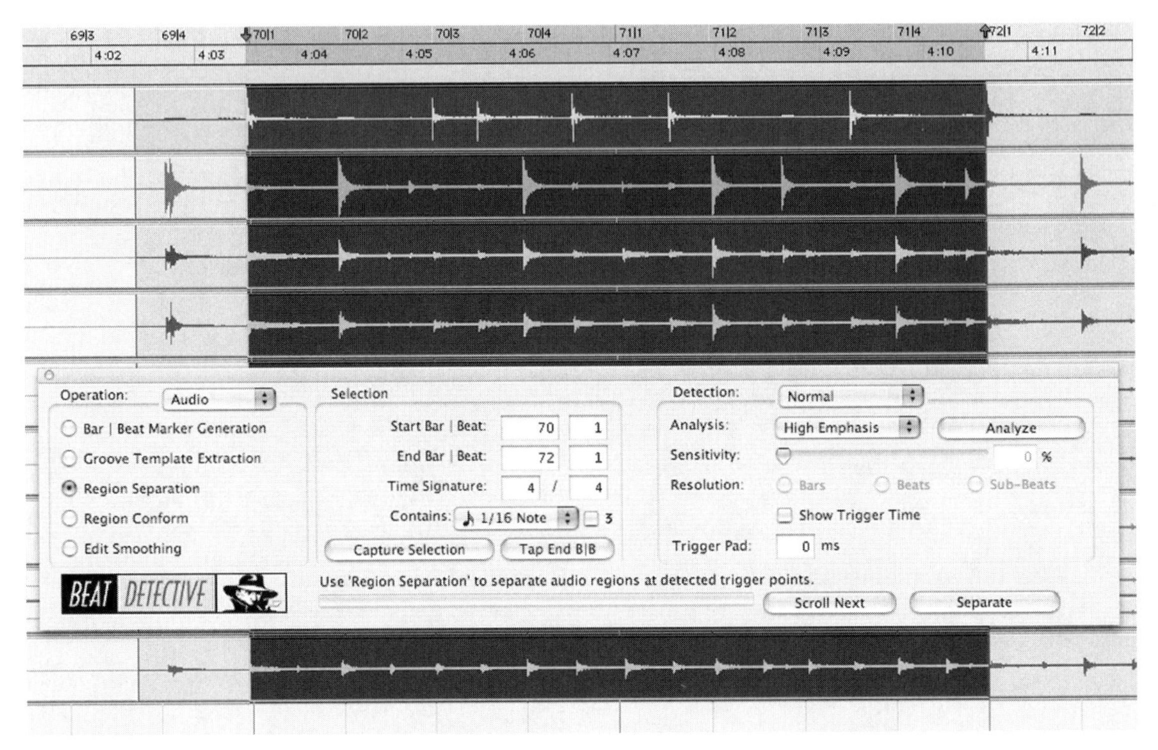

FIGURE 2.11
The two-measure region, ready to be analyzed and separated.

Choosing the detection method

On the right side of the Beat Detective window, you will see three potential options under the "Analysis" selection: "High Emphasis," "Low Emphasis," and "Enhanced Resolution." Selecting either high or low emphasis will focus on either high or low frequencies, respectively. "Enhanced Resolution," found only in Pro Tools 7.4 and higher, has a more complex method of analysis and should be used as your default setting when editing drum tracks with Beat Detective.

Analyzing the audio

Once you have the settings and selection that you need when using Beat Detective, the software will need to analyze the audio, and then you can adjust the sensitivity and resolution accordingly. This is done by pressing the "Analyze" button, and the previously grayed-out sensitivity bar and resolution buttons will become active.

Adjusting the sensitivity for the material

You have the choice in adjusting the sensitivity of Beat Detective by selecting the appropriate option under the detection menu. The sloppier the performance, the higher you are going to want your resolution to be. There are three choices for setting the resolution of Beat Detective.

1. The first is using "Bars," which will place a thick indicator, called a beat trigger, over the selected audio-denoting bars. If you are looking to just tighten up the downbeat of each measure, without affecting the beats in the middle of the bar, then select "Bars."
2. The second choice is "Beats," which will allow Beat Detective to add quarter notes, in addition to bars. These are denoted by a thinner line than the bar triggers, which makes it easier to visualize where Beat Detective will separate the regions.
3. The third choice for resolution is "Sub-Beats," which will then include 8th notes through 16th notes, depending on the selection you have chosen from the "Contains" menu.

After you have selected the appropriate resolution for your selection, you can bring the sensitivity slider up until you see all of the beats in the performance selected with a beat trigger. If you are not seeing all of the beats indicated by Beat Detective, then you may need to choose a smaller selection.

If you have the "Show Trigger Time" button selected, you will see what beats that Beat Detective is identifying the transients to be. These numbers are in the form of Bars | Beats | Sub-Beats, with the Sub-Beats being shown as Pro Tools' MIDI resolution. This MIDI resolution indicates what type of sub-beat is being identified on a scale of 960 ticks per quarter note. If Beat Detective is showing 480 ticks, then it is indicating that it identifies that sub-beat as being an 8th note. A display of 720 ticks would indicate that Beat Detective is identifying that beat as being the fourth 16th note of that particular beat.

Accurately selecting the sensitivity

In general, the lowest that you can bring the sensitivity slider up to where you can see all of the bars, the more accurate it will be. There may be a point when you keep raising the sensitivity, and you will see all of the bars and beats jump to an inaccurate portion of the performance. Increasing the sensitivity will detect softer transients such as hi-hat hits.

In Figure 2.12, you can see that with a sensitivity setting of 19 percent, all of the beats, bars, and sub-beats have been identified. If you look at Figure 2.13, you will see that raising the sensitivity higher, up to 71 percent, indicates that it is inaccurately picking up a 16th note in the decay after the downbeat of the second bar. This can create an audible glitch to the sound after everything has been snapped to its new accurate location.

At this point, it is good to play back the selection to make sure that the beats that you are hearing in the drum parts are indicated by Beat Detective before separating the regions. If everything sounds correct, then press "Separate," and Pro Tools will put edits in all of the regions that it has indicated, and it will adjust them to the correct timing.

Adjusting the beat triggers

The beat triggers that Beat Detective places in the selected region, once the sensitivity has been raised, can be edited. If there are false triggers placed in the selection, those can be deleted by using the Grabber tool and Alt-clicking (PC)

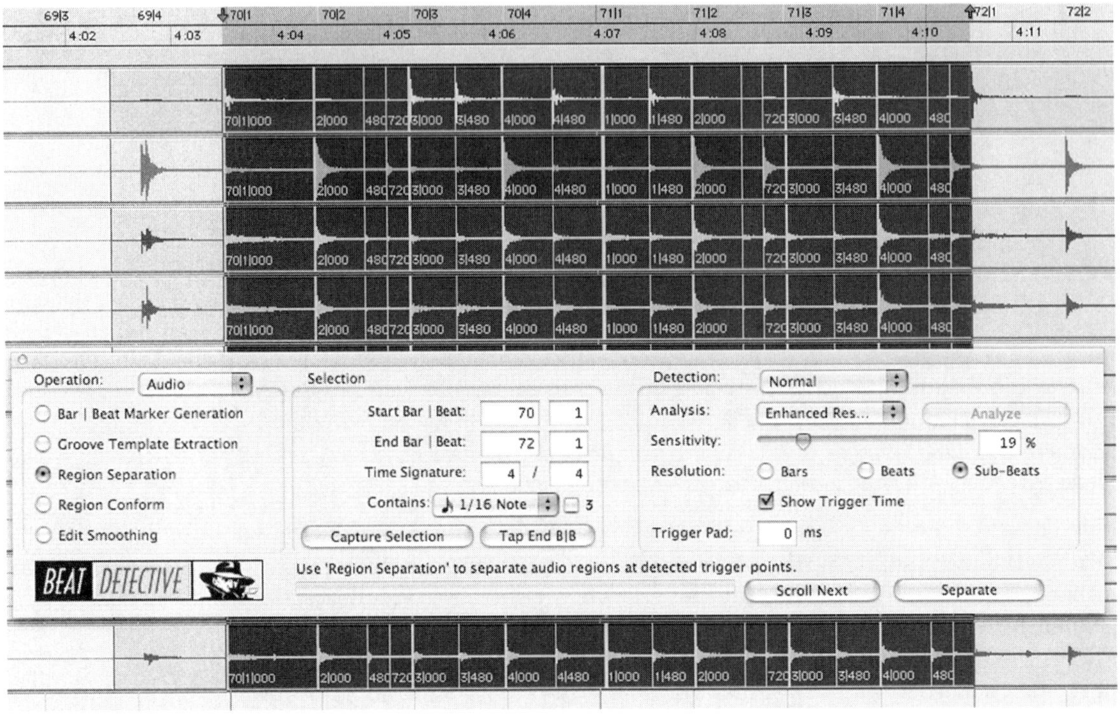

FIGURE 2.12
The region separation markers accurately detecting the beats.

or Option-clicking (Mac) them. In turn, if the transients are not detected as accurately as you may want, you can move the beat triggers by using the Grabber tool and dragging them left or right to the beginning of the transient.

Conforming the regions

After Beat Detective has placed edit points on all of the beats that you are wishing to correct, it is then time to move on to the Region Conform operation by clicking on that button on the left screen (Figure 2.14). Here we have a new set of options on the right side of Beat Detective's window. With "Conform" set to "Standard," which is the correct setting for timing correction, you will see three sliders.

DETERMINING HOW MUCH TO CONFORM

If you want precise accuracy to the drum parts, check the "Strength" box and slide it all the way up to 100 percent, leaving the "Exclude Within" and "Swing" boxes unchecked (Figure 2.15). This will move the separated regions precisely to the tempo map. If you want to maintain some of the tempo fluctuations, you can lower the strength as you see fit.

MAINTAINING SOME OF THE PERFORMANCE

It should be noted that, if you use a setting lower than 100 percent, any sequencing will follow the tempo map and not the drummer's performance. If you want

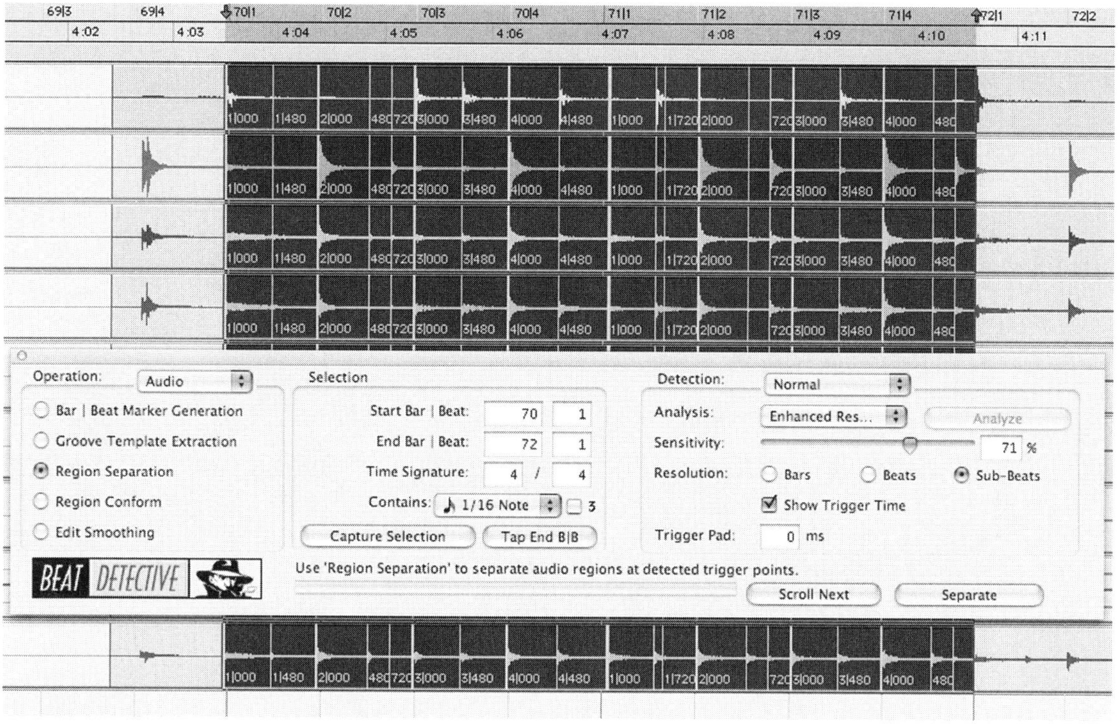

FIGURE 2.13
The region markers have a few noticeable inaccurate detections, due to an increased sensitivity.

to maintain some of the imperfections and still have loops follow the drummer's performance, you will need to adjust the tempo map according to the performance after all of the timing corrections have been made. This is explained in detail in Chapter 1.

THE FINAL CONFORMATION OF THE REGIONS

When you are conforming the selections, the edited selections are moved to their new precisely timed location. In Figure 2.16, you will see that the regions have been conformed. This is a good time to listen to the drum parts with the click track added to the mix so that you can hear if everything has been conformed accurately. You will hear the sounds cut out in the spaces where Beat Detective has created gaps by shuffling the regions around. Do not worry about this, as this will be cleared up in the "Region Conform" stage of using Beat Detective.

FIGURE 2.14
Selecting the Region Conform mode of operation.

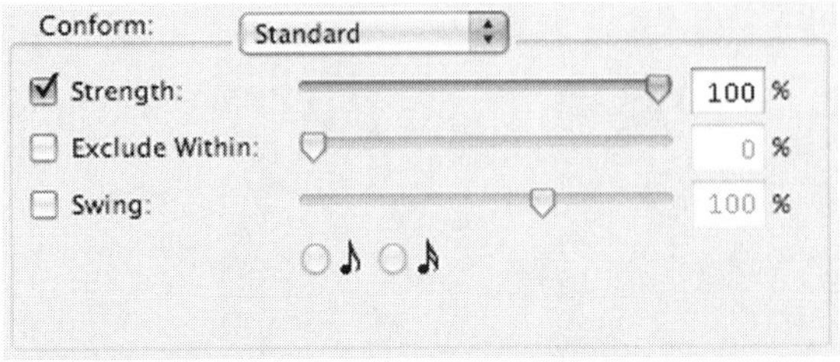

FIGURE 2.15
The conforming strength set to 100 percent for maximum accuracy.

CONTINUING ON WITH BEAT DETECTIVE

Now that you have successfully edited the drum parts for a small section of the track, move on to the next small section throughout the piece. Do not use the "Edit Smoothing" feature until all of the regions have been separated and conformed.

After you have edited this first section, the remaining sections are much easier, as Beat Detective keeps the settings of "Region Separation" and "Region Conform." You may need to make slight adjustments to the sensitivity slider and resolution in a few places, but most similar sections should be easily edited.

SELECTING THE NEXT SET OF BARS

Since we separated the beginning and end of the first Beat Detective regions, the end region is now the first region edit of our next section. We just need to place a region separation at our next chosen end point and continue on, using the same process that we did in the first section.

After the first region separation and conforming, the process will go much faster. After selecting the end point for the next section, double-click in this new section, so that it is all highlighted in Pro Tools, and click "Capture Selection" in

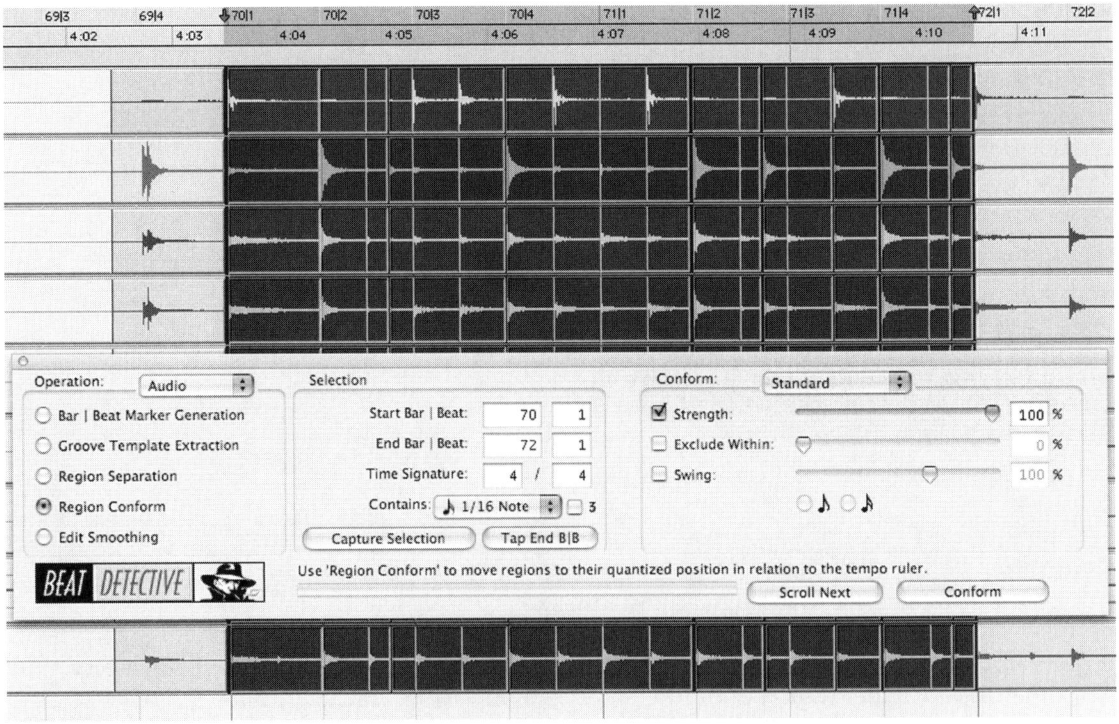

FIGURE 2.16
The separated regions have been conformed at 100 percent strength, with gaps between the adjusted regions.

Beat Detective. This should have adjusted the Start Bar | Beat and End Bar | Beat appropriately, but double check to make sure.

Continue on with the separation and conforming as you did before until you reach the end of the drum tracks that you are using Beat Detective on.

A NOTE ABOUT CONFORMING

Sometimes when conforming regions with Beat Detective, it will shift the last region to the right, which will cover up the transient of the downbeat of the next bar. This happens when the drummer is ahead of the beat. This is easily rectified by using the Trim tool on the unedited track and trimming the unedited region on the right to the left until you see the transient of the downbeat for the next bar.

As you can see in Figure 2.17, the transient downbeat is covered up by the conformed region at the beginning of bar 74. In Figure 2.18, the transient on the right has been brought back by trimming the region on the right over until it has reappeared.

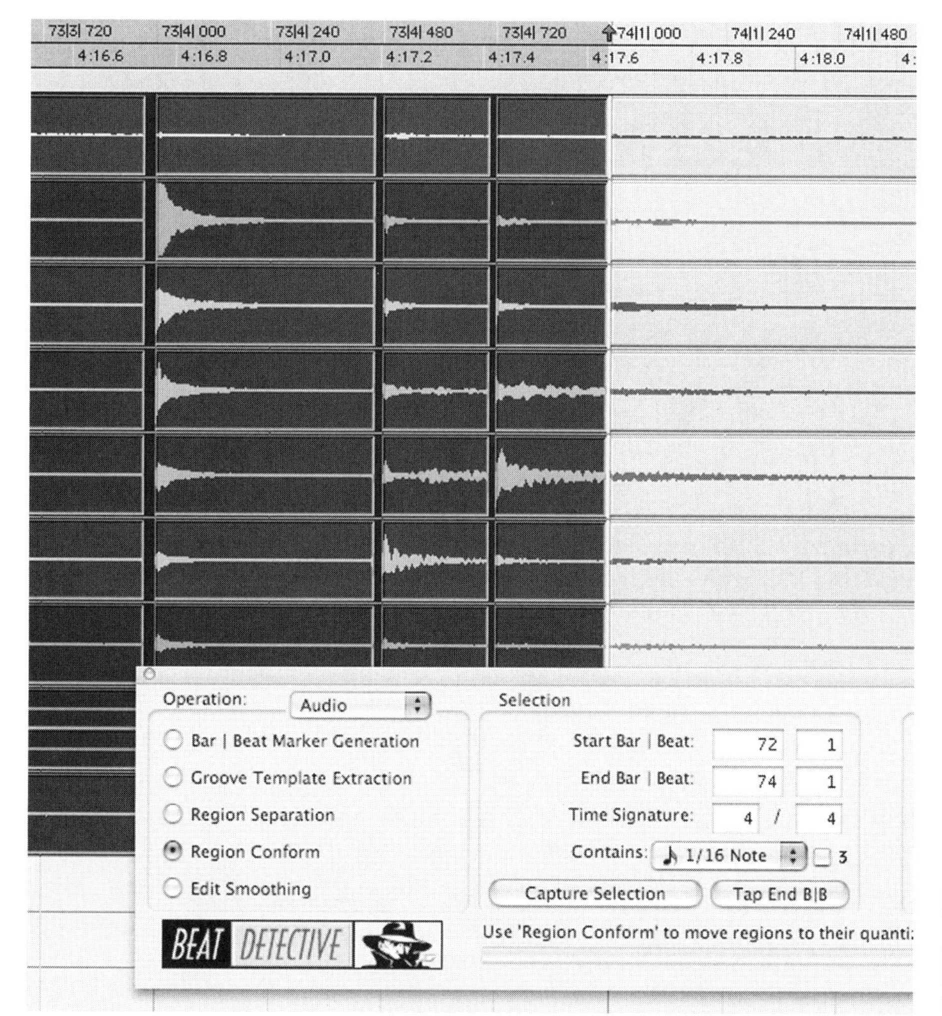

FIGURE 2.17
The transient for the downbeat of bar 74 has been covered up by the conformed regions.

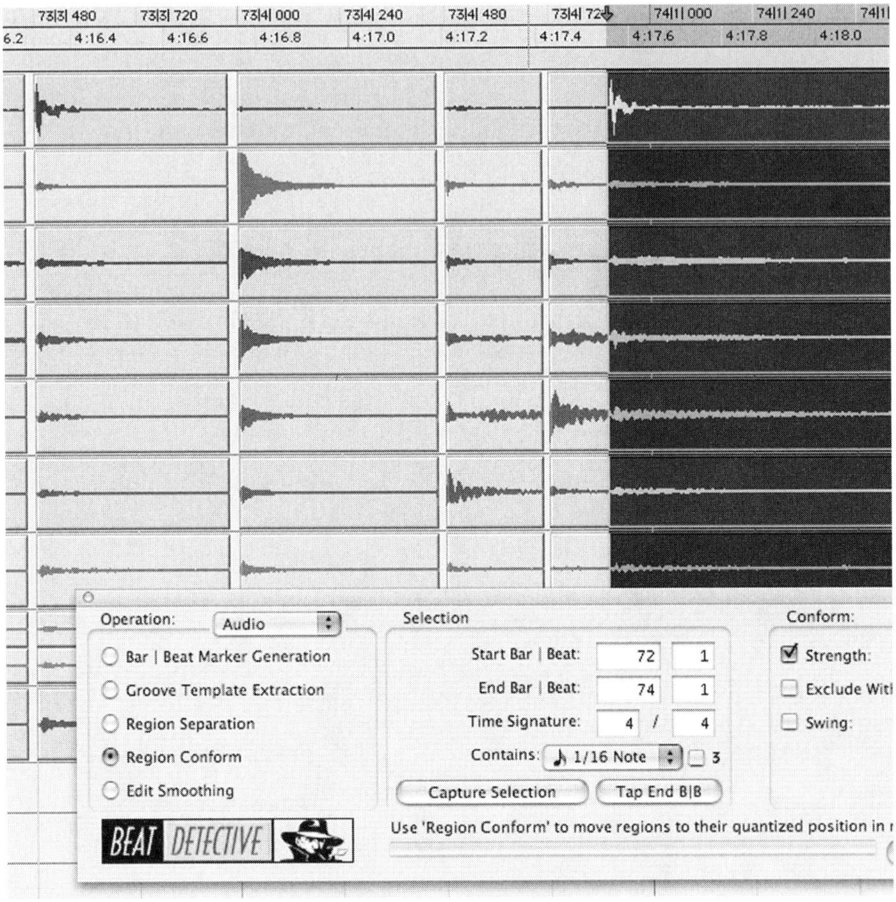

FIGURE 2.18
The transient for bar 74 has been revealed using the Trim tool to drag the start earlier.

LISTENING TO THE NEWLY CONFORMED TRACK

After you have run Beat Detective across the entire track, make sure that you have listened to it to make sure that everything has been adjusted accurately. If there is a bar that does not sound quite right, you may need to individually correct the timing of that bar. This is where the duplicated playlist comes in handy. You can easily paste an unedited bar into the newly conformed playlist.

In order to do this, select the correct playlist number, which will then select all the tracks in the group and adjust them to that playlist as well. Next, highlight the bar that you want to paste over the newly conformed track, and go to "Edit and Copy." This will put the selection into the clipboard. Be sure you do not click anywhere else in the timeline after this.

Select the playlist with the conformed tracks. You will see that the same section has been highlighted. You can just go to "Edit and Paste," and that portion of the original playlist will fall onto the new one.

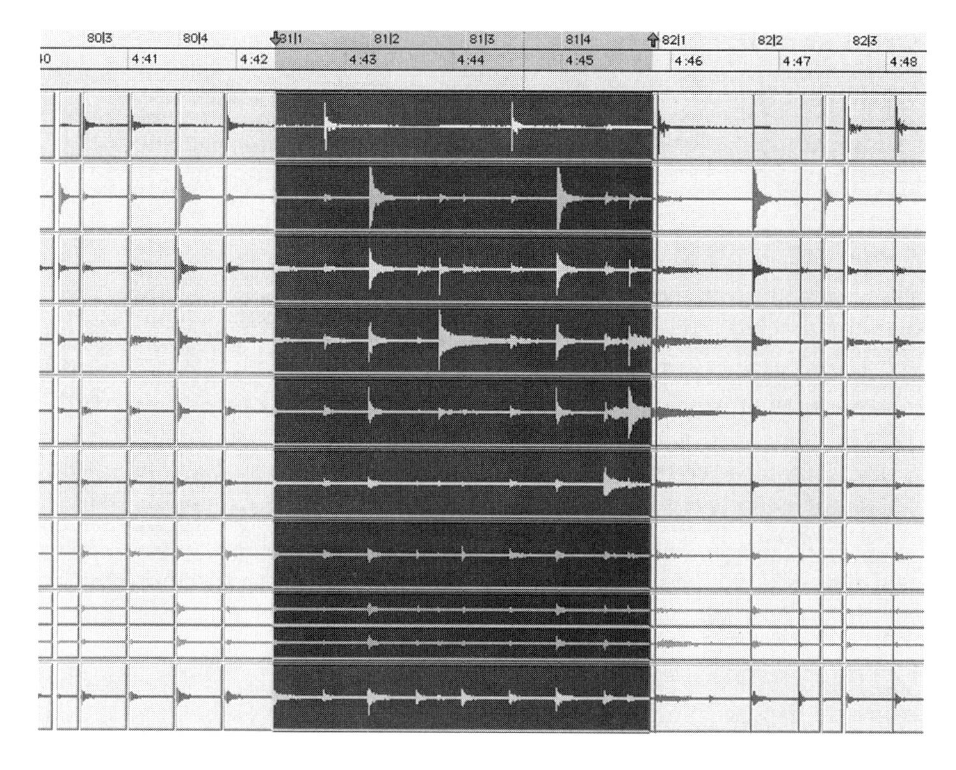

FIGURE 2.19
A previous unedited playlist has been pasted on top of a section previously edited with Beat Detective.

In Figure 2.19, you can see the unedited version pasted into the conformed version. This gives you the ability to always go back to an unedited version if you ever change your mind in the future about having used Beat Detective to do the edits.

Edit smoothing

Before you move on to any edit smoothing, this is another good time to duplicate the playlist of the conformed tracks, so now you have three versions of the drum tracks: an unedited version, a conformed version, and a smoothed version.

Edit smoothing consists of only two choices: "Fill Gaps" and "Fill and Crossfade." When selecting "Fill and Crossfade," you will also have the option to adjust the crossfade length in milliseconds.

FIGURE 2.20
Selecting "Fill and Crossfade" for the edit smoothing operation.

SELECTING THE ENTIRE REGION

Make sure that you have not put any crossfades in the middle of the conformed tracks. Beat Detective will not perform its smoothing operation if there are any crossfades, because the smoothing operation will adjust the boundaries of the regions, and crossfades have no additional boundaries to adjust.

"FILL GAPS" VERSUS "FILL AND CROSSFADE"

Selecting "Fill Gaps" will adjust the boundaries of all of the regions so that they touch the next adjacent region. This will eliminate the dropouts that you hear when there are no regions in between some of the transients, due to the audio that has been moved.

The second selection of "Fill and Crossfade" will additionally place a short crossfade with the length of your choosing at each of these region edits. This can create issues when transients are crossfaded into another transient, creating a false hit, so be sure that you listen to the track right after you have smoothed them.

The first thing you can do is to try and fill gaps without crossfading them. This will give you the ability to make sure that there are no false double transients. If the drummer's performance is way off, there is the possibility of these false transients showing up in the smoothed version that you did not hear in the conformed version. You can eliminate these false transients by manually adjusting the edit with the Trim tool.

After you fill the gaps, you can optionally try to crossfade them as well, using the "Fill and Crossfade" feature, as it will only add the crossfades after the selection has already been filled.

FILLING THE GAPS

Filling the gaps is a quick process. Make sure that you listen to the tracks to confirm that there is nothing wrong with the adjusted regions.

In Figure 2.21, you can see the drum tracks that have not been smoothed. In Figure 2.22, you will see that the gaps between the regions have been smoothed by having their boundaries adjusted.

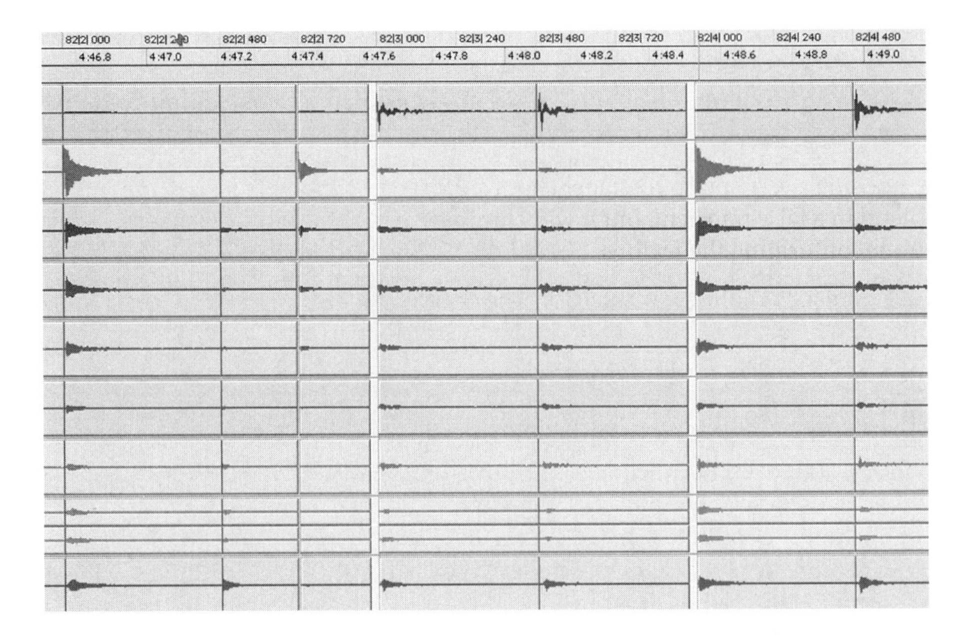

FIGURE 2.21
Regions separated and conformed, but not yet smoothed.

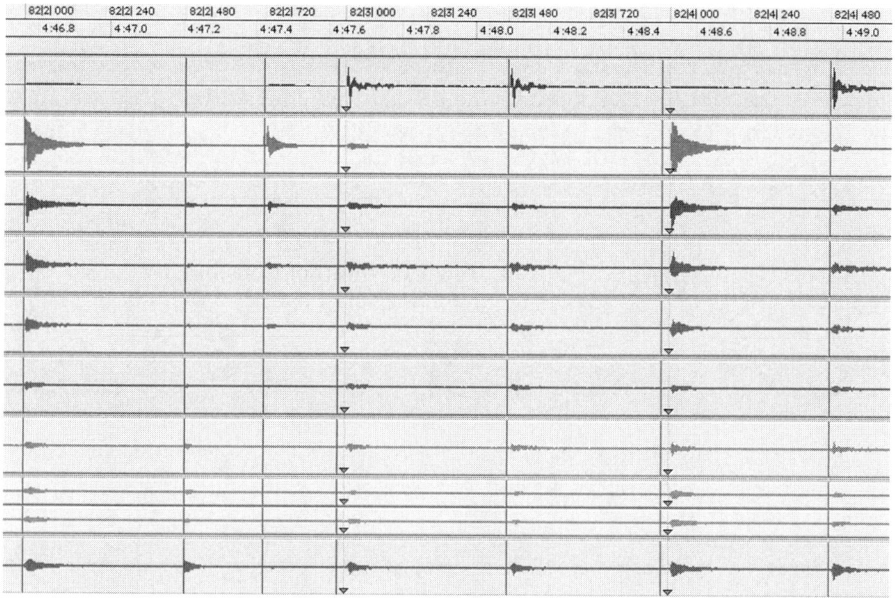

FIGURE 2.22
The conformed regions have been smoothed, eliminating any gaps between regions.

CROSSFADES

Adding a crossfade to the selection actually writes a new audio file that consists of that crossfaded region. This saves processing by eliminating the need for the software to perform the crossfading in real time and takes up only a little bit of hard disk space. Crossfading can eliminate any pops and clicks that you may hear due to non-zero-crossing edits of the audio.

Choosing to use the crossfade feature requires listening to the sound to see if it smoothes the sound out better, or if it is creating more artifacts, with the added bonus of making the sound smoother. If you are using the crossfade, you have the option of pasting in an unsmoothed version from the previously used playlist in order to eliminate any artifacts.

Choosing a very small crossfade length such as 1 ms will have less of a chance to blend in a false transient, but it will eliminate any pops and clicks that may arise from conforming the regions.

In Figure 2.23 of the same selection, the crossfade feature was added to the same previous selection. You can see the very small crossfades that are over each of the edit points.

Consolidating the edited tracks

Once you have run all the phases of Beat Detective and there are no artifacts to the audio, you can choose to consolidate the selection. Before you do this, be sure that you duplicate the playlist so that you can always go back to this version in case you missed an added false transient, or the track is not smoothed the way that you want.

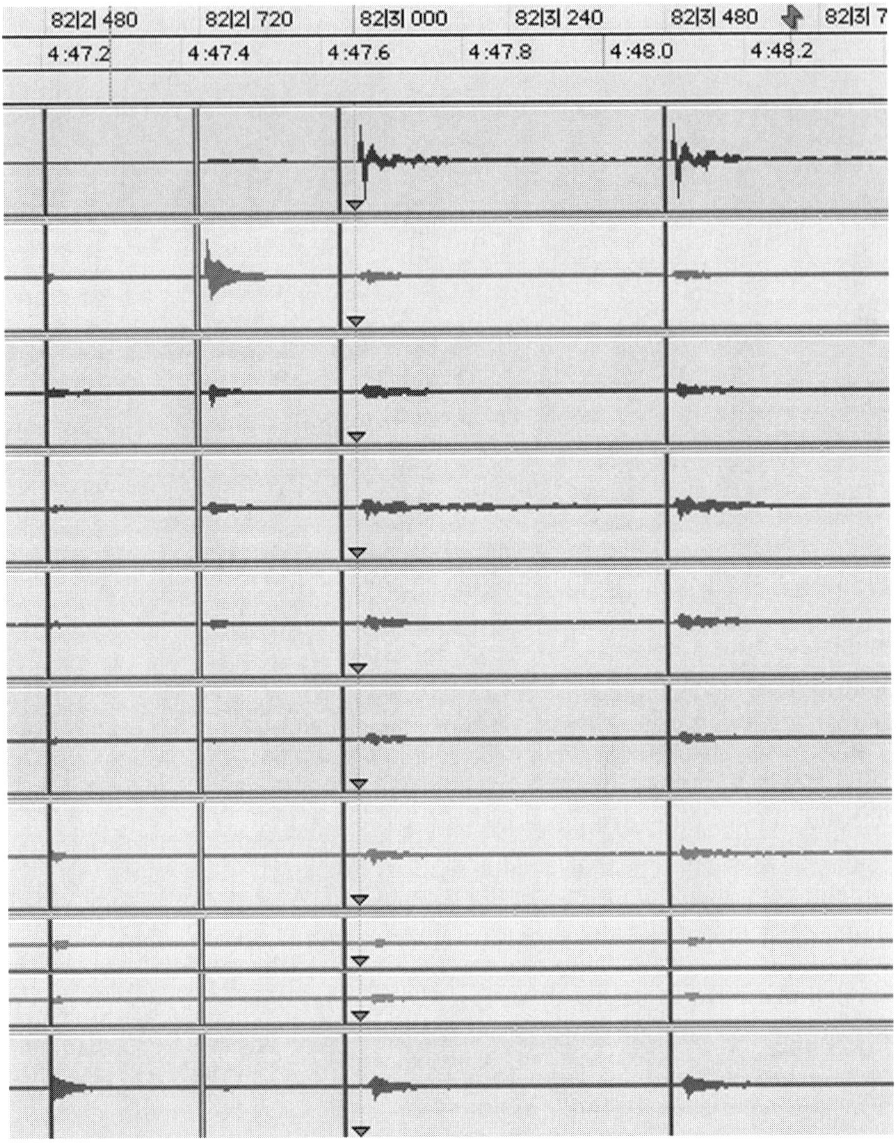

FIGURE 2.23
The conformed regions with a 1-ms crossfade added in between.

Consolidating the tracks rewrites all the selected audio into a continuous audio file. Since Beat Detective generates hundreds of edits and crossfades, there are many files that Pro Tools is trying to read in real time as you play through the track. This will take up processing as well as bandwidth reading from your hard drive. Consolidation is the solution to this issue.

In order to consolidate the tracks, highlight all the contents from beginning to end. Go to the Edit menu and select "Consolidate," and Pro Tools will begin consolidating all of the highlighted audio regions. Since this is writing new audio tracks, it will take up more hard disk space and will add additional files to your Audio Files window.

Speeding up the beat with Beat Detective

One of the more interesting tricks that you can do with Beat Detective is to use it to adjust the speed of a performance.

An easy thing to do is to take a two-measure selection and convince Beat Detective that it is a one-measure selection. In this example, we will use the first two measures that we originally applied Beat Detective to. When going to the Region Separation operation, tell Beat Detective that the End Bar | Beat is bar 71, even though you have through bar 72 highlighted.

In Figure 2.24, you will see that the "Contains" has been adjusted so that it halves the length of the previous resolution from 16th notes to 32nd notes. This is because we are going to trick Beat Detective into thinking that these two bars are actually one bar.

Go through the analysis and separation like you did before. You may need to increase the sensitivity and do some manual editing of the beat triggers if you are going to try and capture the softer hi-hat notes. Then conform the selection as you have done previously.

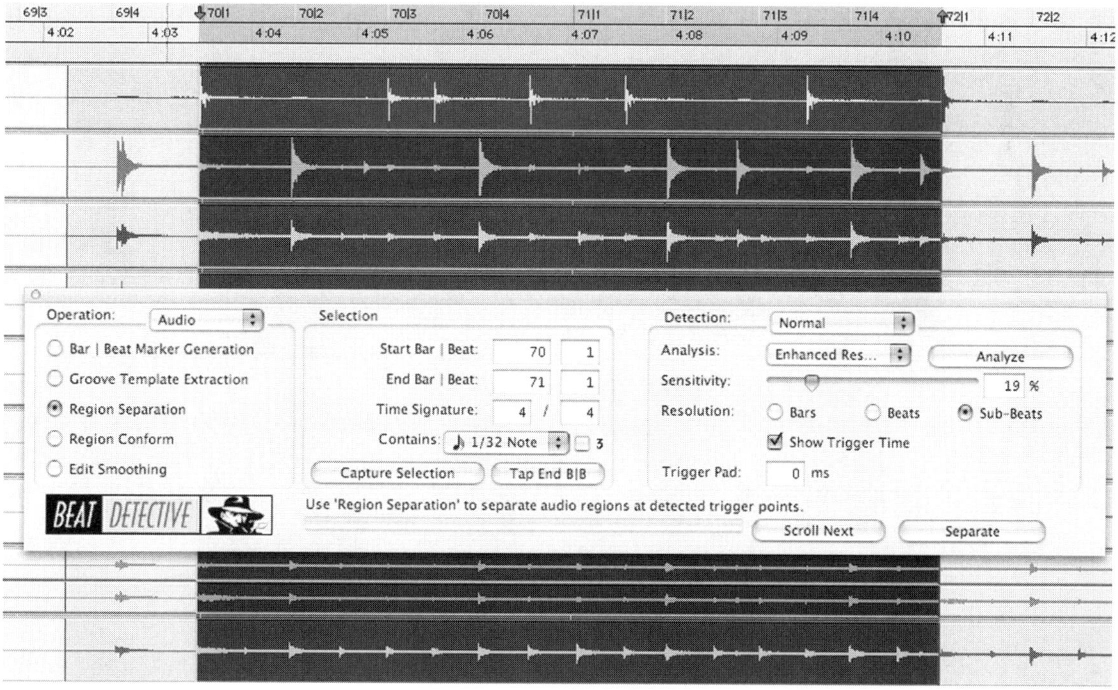

FIGURE 2.24
Forcing Beat Detective to recognize a two-measure selection as one measure.

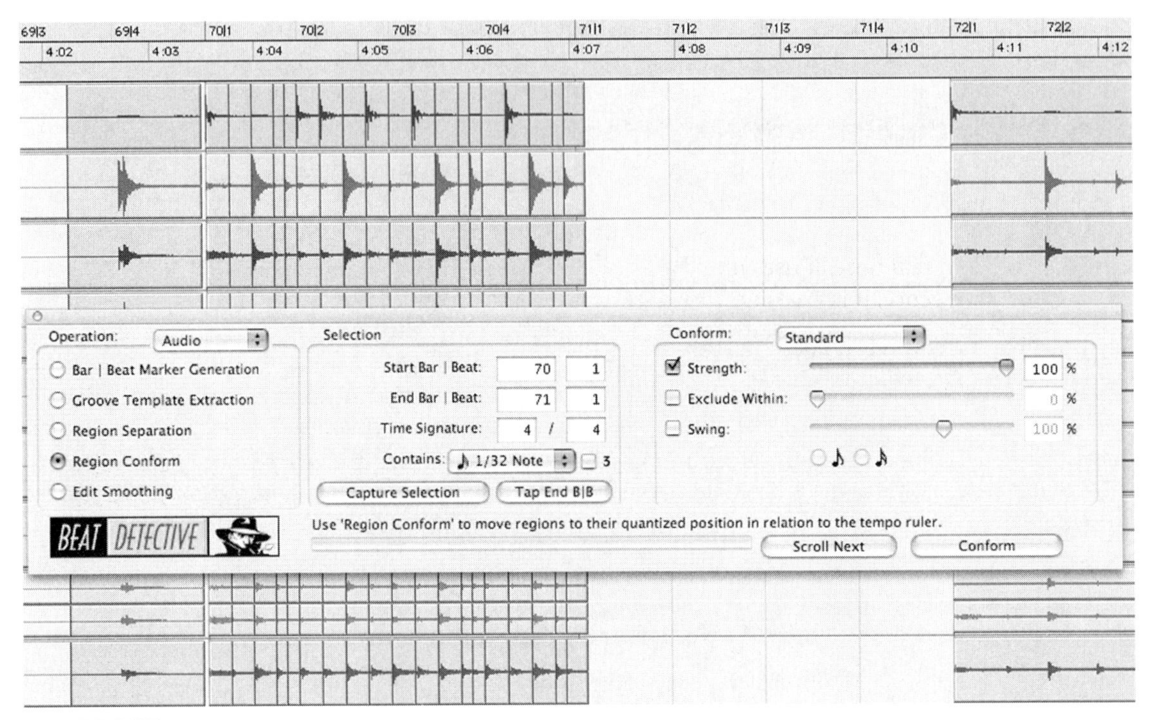

FIGURE 2.25
The two-measure selection has been separated and conformed onto one measure.

In Figure 2.25, you will see that now we have the time-compressed version, but rather than using a time-compression algorithm, Beat Detective has simply doubled the performance through the use of edits to the regions.

This can be an interesting way to accomplish time compression. The sounds can come across as being gated, as the decays are being cut off with each subsequent edit in an unnatural way; however, this can be one of the more interesting sound design methods in your toolbox.

ELASTIC TIME

Elastic Time is one of the main features of Digidesign's Pro Tools 7.4. It allows for the manipulation of time and pitch through the use of various selectable algorithms. It has some of the same functionality as the time-compression and -expansion algorithms; however, it works in real time, which allows for nondestructive adjustments to be made in the audio tracks.

The two most powerful editing functions of Elastic Time that can be used in record production are the alignment of timing of vocals or other pitched instruments, and the correction of the drum and percussion tracks. It may seem redundant to use Elastic Time to correct the drums when you have the use of Beat Detective, but it merely becomes a different tool at your disposal.

Beat Detective works differently than Elastic Time. Beat Detective automatically splices and aligns audio based on the transients detected. Elastic Time is able to stretch and compress the audio based on detected transients. This has the advantage of allowing you to not worry about crossfades and extra transients being created through the smoothing process in Beat Detective.

Elastic Time has different means of analyzing audio. In order to make the most out of Elastic Time, selecting the appropriate algorithm is important. Since Elastic Time functions in real time, it uses up processing for each of the tracks on which it is applied. In essence, it is an optional real-time plug-in with the option of rendering the performance as audio files on your hard drive so that you are not taxing your computer processing unit (CPU).

FIGURE 2.26 Selecting the different algorithms for Elastic Time from the Edit window.

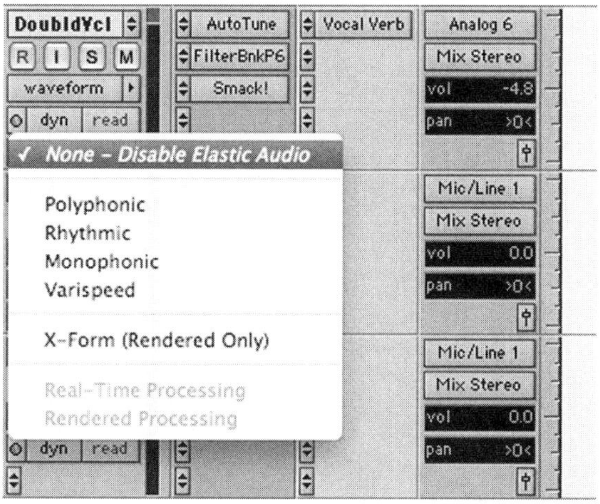

There are three main algorithms in Elastic Time that are used to correct and adjust timing: polyphonic, rhythmic, and monophonic. There is also a varispeed algorithm that can create different effects on the audio by speeding up and slowing down the audio based on the tempo changes of the tempo map in relation to the warp markers.

Engaging Elastic Time is as simple as clicking a selector tab underneath the track's name at the bottom of each track in the edit window. In order to best use Elastic Time, you need to select the algorithm that is best suited for your material (Figure 2.26).

Elastic Time algorithms

POLYPHONIC MODE

The polyphonic algorithm is best suited for tracks that have more than one note playing simultaneously. This algorithm is useful on guitar or keyboard tracks.

RHYTHMIC MODE

The rhythmic algorithm is best suited for drums and percussion. It will keep the transient attacks intact while still being able to manipulate and stretch the audio. This algorithm would not be the best choice for any specifically pitched audio.

MONOPHONIC MODE

The monophonic algorithm will work on a track that has only one note playing at a time. This is best suited for lead and background vocals or individual horn parts.

VARISPEED MODE

The varispeed algorithm will create special effects to the audio. When the audio is being stretched, Elastic Time will slow down the audio track as if it were an analog tape machine. When the audio is being compressed, the sound will be pitched up.

Analyzing the audio track

Elastic Time, once engaged by selecting an algorithm, will then analyze the audio on the track. If there are multiple tracks in a group it may take some time for Elastic Time to complete its analysis. The track will be grayed out until the analysis is complete.

This analysis does not analyze the frequency content of the audio track, rather it merely analyzes the track for transients. This is used if you are specifically going to quantize the track to the tempo map.

WARP VIEW

Once an algorithm has been selected and the audio track has been analyzed, you now have the option of switching to view the track in warp and analysis view (Figure 2.27). Looking at the track in warp view, you now have the ability to insert warp markers, and can now go about stretching and compressing the audio.

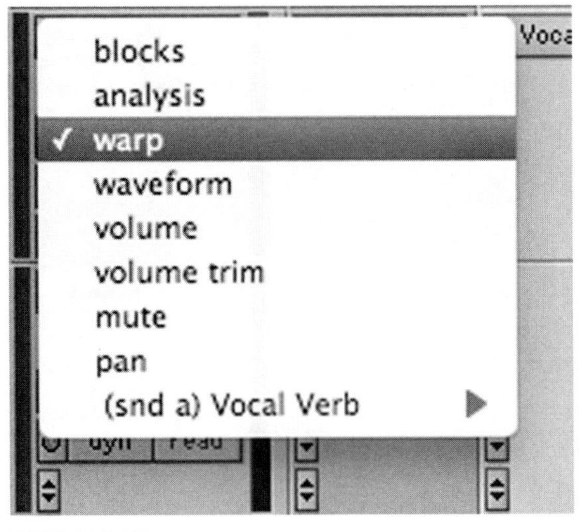

FIGURE 2.27
Adjusting the view of the track to create and edit warp markers.

Aligning vocals using Elastic Time

Anytime you are recording with multiple vocal tracks, there may be a time in which you will need to align either background vocals or a doubled vocal part to the lead vocal. Elastic Time makes a great tool for manually adjusting the timing of these vocal parts to the lead vocal.

In order to align vocal parts using Elastic Time you will need to insert warp markers. If you are editing a track in the warp view, you can either select the Pencil tool and add them in manually or use the Grabber tool and Start-click (PC) or Control-click (Mac). It is easiest to use the Grabber tool, as this is the same tool that is used to drag the warp markers in warp view. Additionally, warp markers can be deleted by simply pressing the Option key and clicking on the warp marker with either the Pencil tool or the Grabber tool.

Elastic Time will stretch and compress any audio on either side of the warp markers. A good habit to get into would be to place a warp marker at the beginning and end of the selection that you are adjusting. This will prevent Elastic Time from compressing or expanding the entire audio track (Figure 2.28).

When using Elastic Time for vocal manipulation, it is best to go and edit the vocals phrase by phrase. Placing a warp marker at the beginning and end of each phrase or note will allow for just the time compression and expansion of those individual phrases (Figure 2.29).

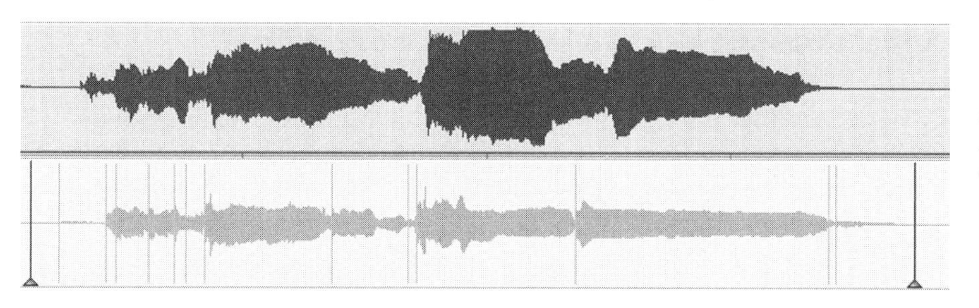

FIGURE 2.28
Inserting warp markers before and after a selection to prevent accidental adjustment of regions to the left and right.

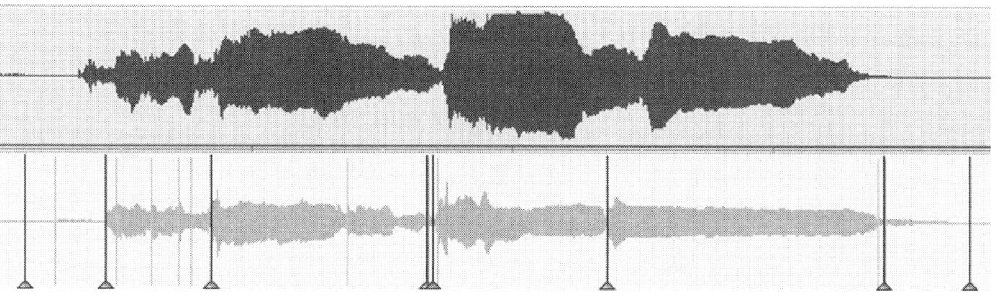

FIGURE 2.29
The warp markers placed at the beginning and ending of each phrase.

After the markers have been placed, select the Grabber tool. From here, it is merely a matter of dragging the markers in such a way that they will line up the beginning and ending of each of the notes or phrases. You may find that you may need to add an extra warp marker in a few locations to get the timing to be exactly what you want. See Figure 2.30.

Using Elastic Time to quantize drum parts

Elastic Time has the ability to effectively quantize drum parts in much the same way that you would quantize MIDI tracks. Pro Tools will analyze the drum parts for transients and then align the transients to the grid based on the resolution that you choose. This works only if the tempo map is accurate to the recorded tracks. If you recorded to a click track from inside Pro Tools, then the tempo map will match up with your editing.

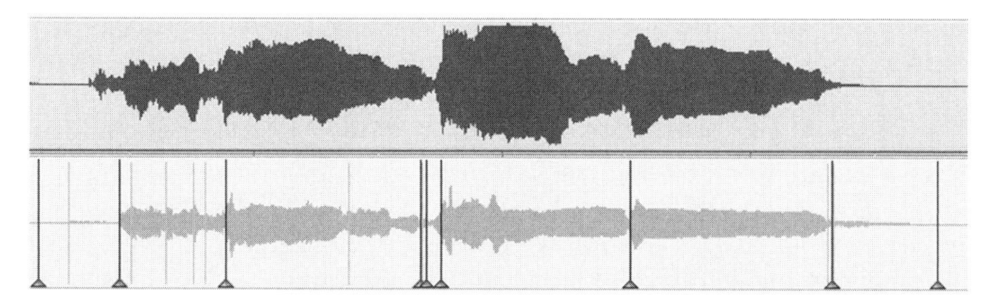

FIGURE 2.30
The warp markers have been adjusted to make the doubled vocal match the original.

Sample-based tracks versus tick-based tracks

There is a difference between working with MIDI tracks and audio tracks in your DAW. MIDI tracks contain only data and no audio, while the audio tracks are all sample based. When making tempo changes in Pro Tools, MIDI tracks will follow the tempo changes while the audio tracks remain fixed in their location in the timeline.

Since Elastic Time has the ability to compress and expand audio tracks, it now has the ability to expand and contract those tracks to follow any tempo changes that you may make. In essence, this means that you can treat audio tracks in the same way as you treat MIDI tracks. You gain the ability to quantize the parts in the same way as you would quantize MIDI data.

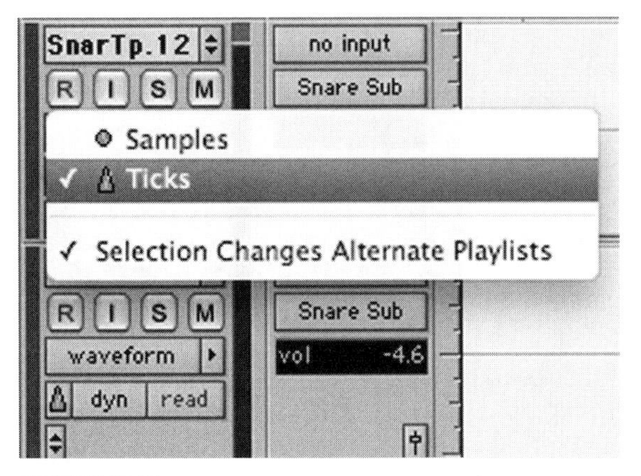

FIGURE 2.31
Selecting the track to go from sample based to tick based.

In order to get the audio tracks to function the same way as MIDI tracks, you need to change the audio tracks from sample-based tracks to tick-based tracks. This will lock the audio data into their bar locations as opposed to their position in the timeline. The audio between transients will then be compressed or expanded to adjust to the tempo changes. To switch a track from sample based to tick based, click on the box below the view in the Edit window of the track and change your selection appropriately (Figure 2.31). With all the drum tracks in a group, you can switch them all at once.

Engaging Elastic Time

To engage Elastic Time, all you need to do is select the rhythmic algorithm in order for it to begin its analysis. Once all the tracks have been analyzed for their transients, the initial quantization of the parts becomes easy. Throughout the course of correcting the time in the drum tracks, you will find it helpful to turn on a click track so that you can make sure that any quantization done is accurate.

QUANTIZING THE DRUM TRACKS AS EVENTS

As with any editing process, duplicate the playlist of the drum tracks so that you can always go back to the starting point. To begin quantizing the drum tracks, start by switching the view to "Analysis." From here you will see that Elastic Time has placed black markers on any of the transients that it detects (Figure 2.32).

Once you have the selection highlighted that you want to quantize, go to the Events menu and select "Event Operations" and then "Quantize" (Figure 2.33). This will treat these tick-based tracks similarly to the way that Pro Tools treats MIDI tracks.

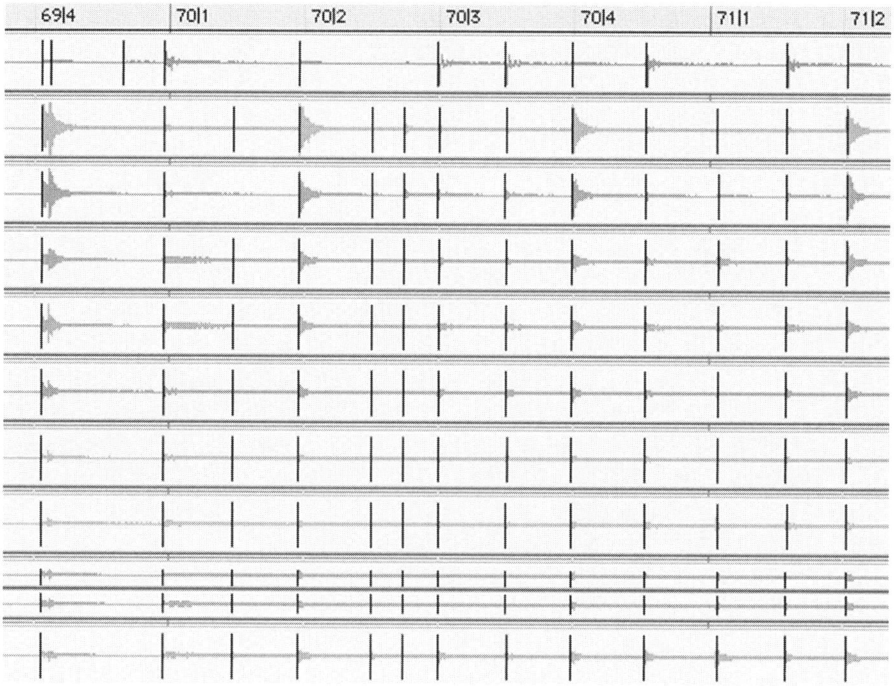

FIGURE 2.32 Transients detected by Elastic Time as black lines.

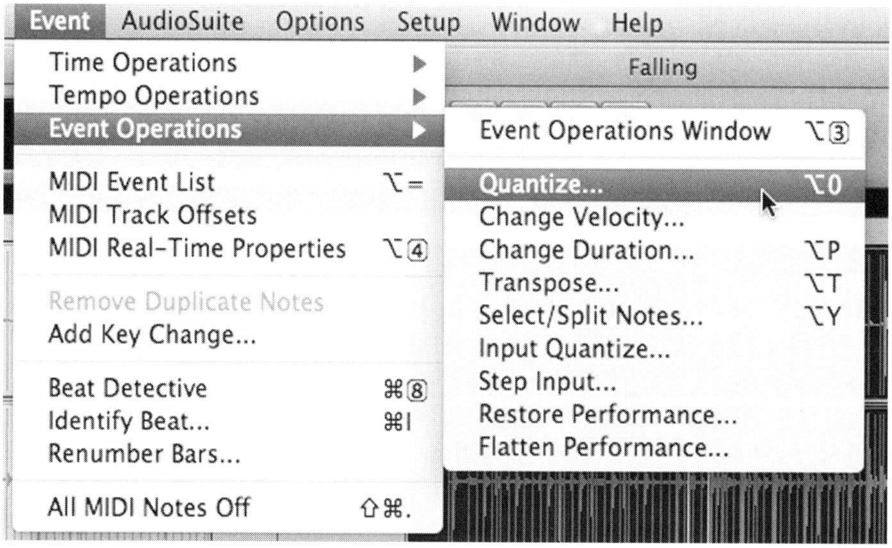

FIGURE 2.33 Selecting "Quantize" from the Events menu.

The "Event Operations" window will then give you the same options that it does if they were MIDI tracks. Select the quantized grid based on the drummer's performance and then the strength of the quantization. You have the option of not quantizing it 100 percent to the tempo map; however, if you want a very precise performance, increase the strength as high as you feel comfortable (Figure 2.34).

FIGURE 2.34
Quantize settings for Elastic Time to snap the audio to the nearest 16th note.

MANUALLY ADJUSTING ANY QUANTIZING MISCALCULATIONS

Once you have applied the quantization across the tracks, Elastic Time places warp markers and adjusts those warp markers according to the tempo map (Figure 2.35). You will find that Elastic Time may not be 100 percent accurate over the entire performance, and you may need to correct a few single hits here or there. Switch to view the tracks in warp, and you can now make the annual adjustments as necessary.

Making corrections to the quantization using Elastic Time is very easy to do. Correcting these errors is merely a matter of manually removing and placing a new warp marker (Figure 2.36). Warp markers can easily be deleted by selecting the Grabber tool and Option-clicking on the misplaced warp marker.

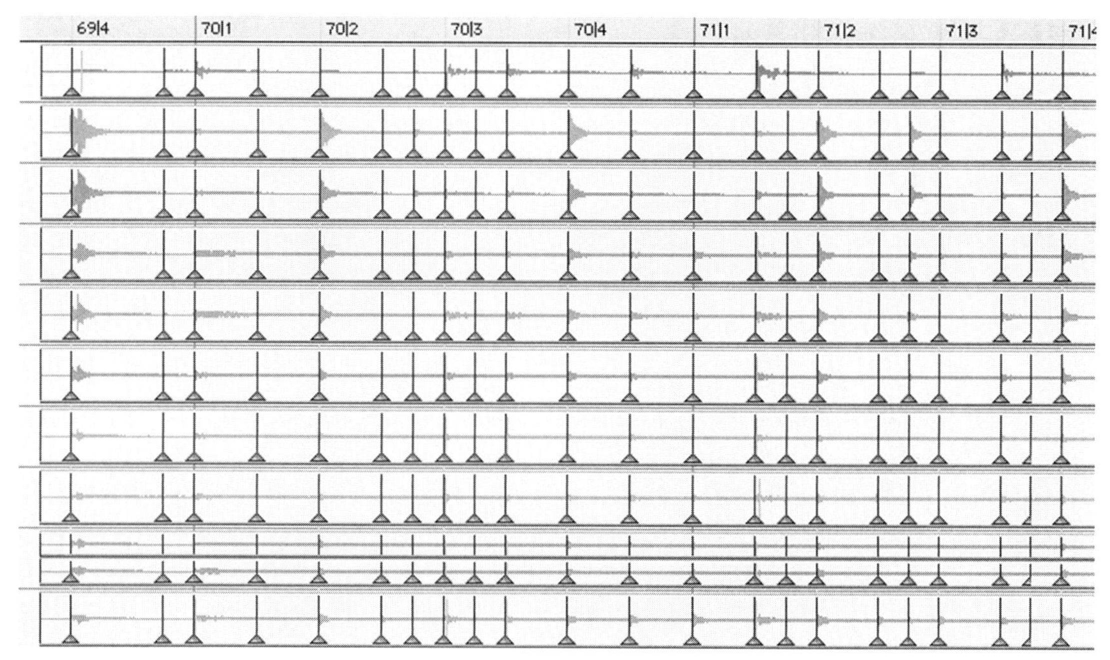

FIGURE 2.35
The warp markers have been moved after quantizing the tracks.

FIGURE 2.36
Warp markers that have missed the transient and need manual adjustment.

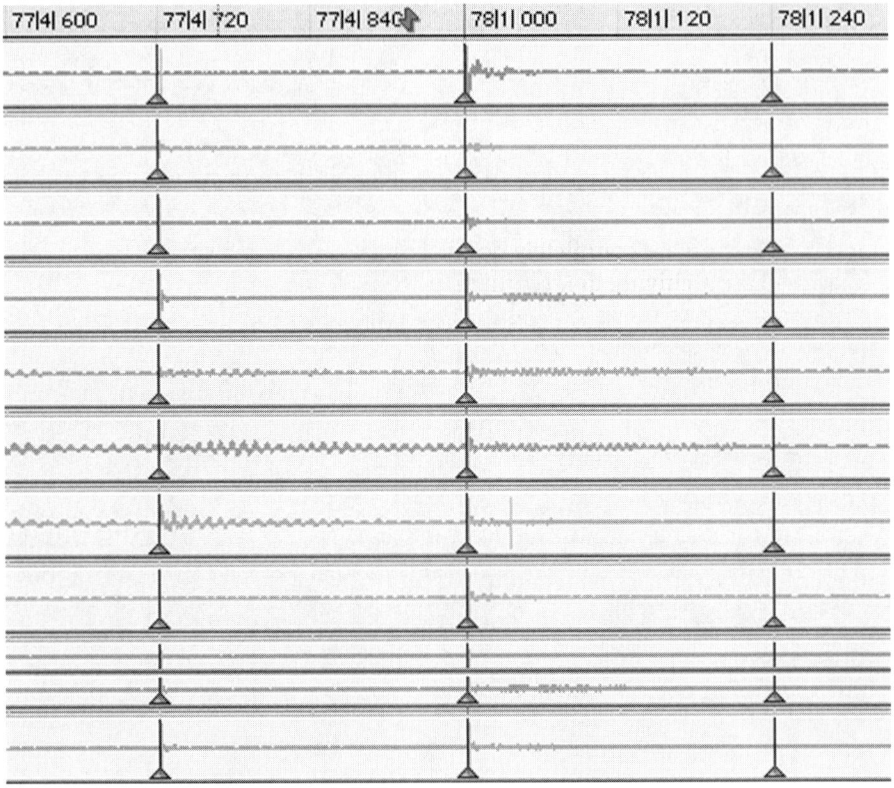

FIGURE 2.37
Warp markers that have been adjusted to be on the kick drum hit, and then manually snapped to the correct beat.

Control-clicking, while still in the Grabber mode, will create the new warp marker that you need to correct the quantization. In order to lock the new hit on time, switch Pro Tools from Slip mode to Grid mode, making sure that the grid is set up to the appropriate resolution of the beats as necessary. This will make the newly created warp marker snap to grid when you drag it to its corrected location (Figure 2.37).

Rendering the audio files

Depending on your computer's processing, you may find that running Elastic Time across several tracks may use up too much of your computer's CPU power. Elastic Time makes it easy to compensate for this by rendering the tracks as opposed to processing them in real time.

Rendering tracks will rewrite the audio files with the warp adjustments. This will add audio files to your hard disk but save in processing (Figure 2.38). Rendering the audio files is best done after you have completed all of your major editing of the warp markers. While in rendered mode, you still have the ability to manipulate warp markers; however, each adjustment you make will cause Pro Tools to rerender each of those tracks. This may not be suited for a situation where you are trying to work quickly, but it is not a big deal if you are only going to need to make a few adjustments to the warp markers.

After you have done several corrections using Elastic Time, if you turn off Elastic Time from the tracks, Pro Tools gives you the option of committing those warp markers to the audio tracks. Unless you are completely done with the project and are archiving the recording, it would be best to keep the tracks in rendered mode so you can still make adjustments easily up until the very end.

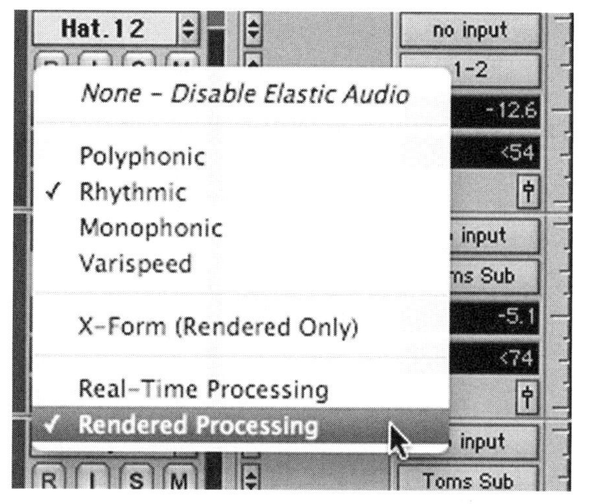

FIGURE 2.38
Selecting "Rendered Processing" to write the audio file and save the computer's CPU.

A FINAL WORD ABOUT TIMING CORRECTION

Correcting any timing issues is important to do before moving on to other production techniques. Since everything will be based on the rhythm of the basic tracks as well as the overdubs, imperfections and timing will have a cumulative effect on the recording. If the overdubbing is taking place on top of mistimed drums or other instruments, it becomes more difficult to recognize whether these overdubbed parts are in time with the track.